The IMA Volumes in Mathematics and Its Applications

Volume 26

Series Editors
Avner Friedman Willard Miller, Jr.

Institute for Mathematics and its Applications
IMA

The **Institute for Mathematics and its Applications** was established by a grant from the National Science Foundation to the University of Minnesota in 1982. The IMA seeks to encourage the development and study of fresh mathematical concepts and questions of concern to the other sciences by bringing together mathematicians and scientists from diverse fields in an atmosphere that will stimulate discussion and collaboration.

The IMA Volumes are intended to involve the broader scientific community in this process.

Avner Friedman, Director
Willard Miller, Jr., Associate Director

* * * * * * * * * *

IMA PROGRAMS

1982-1983	**Statistical and Continuum Approaches to Phase Transition**
1983-1984	**Mathematical Models for the Economics of Decentralized Resource Allocation**
1984-1985	**Continuum Physics and Partial Differential Equations**
1985-1986	**Stochastic Differential Equations and Their Applications**
1986-1987	**Scientific Computation**
1987-1988	**Applied Combinatorics**
1988-1989	**Nonlinear Waves**
1989-1990	**Dynamical Systems and Their Applications**
1990-1991	**Phase Transitions and Free Boundaries**

* * * * * * * * * *

SPRINGER LECTURE NOTES FROM THE IMA:

The Mathematics and Physics of Disordered Media
Editors: Barry Hughes and Barry Ninham
(Lecture Notes in Math., Volume 1035, 1983)

Orienting Polymers
Editor: J.L. Ericksen
(Lecture Notes in Math., Volume 1063, 1984)

New Perspectives in Thermodynamics
Editor: James Serrin
(Springer-Verlag, 1986)

Models of Economic Dynamics
Editor: Hugo Sonnenschein
(Lecture Notes in Econ., Volume 264, 1986)

Daniel D. Joseph David G. Schaeffer
Editors

Two Phase Flows and Waves

With 48 Illustrations

Springer-Verlag
New York Berlin Heidelberg
London Paris Tokyo Hong Kong

Daniel D. Joseph
Department of Aerospace
Engineering and Mechanics
University of Minnesota
Minneapolis, MN 55455
USA

David G. Schaeffer
Department of Mathematics
Duke University
Durham, NC 27706
USA

Series Editors

Avner Friedman
Willard Miller, Jr.
Institute for Mathematics and Its Applications
University of Minnesota
Minneapolis, MN 55455
USA

Mathematics Subject Classification: 00-02, 00-11, 35Lxx, 35L65, 35L67, 35L70, 35M05, 70-XX, 70-02, 70-05, 70-06, 70-08, 70A05, 70Dxx, 70Kxx, 70K20, 70Q05, 76Bxx, 76D33, 76T05

Library of Congress Cataloging-in-Publication Data
Two phase flows and waves / [edited by] Daniel D. Joseph, David G. Schaeffer.
p. cm. — (The IMA volumes in mathematics and its applications ; v. 26)
Papers based on a symposium held Jan. 3-10, 1989 at the University of Minnesota Institute for Mathematics and Its Applications.
ISBN-13:978-1-4613-9024-4
1. Two-phase flow—Congresses. 2. Wave-motion, Theory of—Congresses. I. Joseph, Daniel D. II. Schaeffer, David G. III. University of Minnesota. Institute for Mathematics and Its Applications. IV. Series.
QA922.T96 1990
532'.051—dc20 90-33861

Printed on acid-free paper.

Camera-ready text prepared from an IMA LaTeX file.

9 8 7 6 5 4 3 2

ISBN-13:978-1-4613-9024-4 e-ISBN-13:978-1-4613-9022-0
DOI: 10.1007/978-1-4613-9022-0

The IMA Volumes
in Mathematics and its Applications

Current Volumes:

Volume 1: Homogenization and Effective Moduli of Materials and Media
Editors: Jerry Ericksen, David Kinderlehrer, Robert Kohn, J.-L. Lions

Volume 2: Oscillation Theory, Computation, and Methods of Compensated Compactness
Editors: Constantine Dafermos, Jerry Ericksen,
David Kinderlehrer, Marshall Slemrod

Volume 3: Metastability and Incompletely Posed Problems
Editors: Stuart Antman, Jerry Ericksen, David Kinderlehrer, Ingo Muller

Volume 4: Dynamical Problems in Continuum Physics
Editors: Jerry Bona, Constantine Dafermos, Jerry Ericksen, David Kinderlehrer

Volume 5: Theory and Applications of Liquid Crystals
Editors: Jerry Ericksen and David Kinderlehrer

Volume 6: Amorphous Polymers and Non-Newtonian Fluids
Editors: Constantine Dafermos, Jerry Ericksen, David Kinderlehrer

Volume 7: Random Media
Editor: George Papanicolaou

Volume 8: Percolation Theory and Ergodic Theory of Infinite Particle Systems
Editor: Harry Kesten

Volume 9: Hydrodynamic Behavior and Interacting Particle Systems
Editor: George Papanicolaou

Volume 10: Stochastic Differential Systems, Stochastic Control Theory and Applications
Editors: Wendell Fleming and Pierre-Louis Lions

Volume 11: Numerical Simulation in Oil Recovery
Editor: Mary Fanett Wheeler

Volume 12: Computational Fluid Dynamics and Reacting Gas Flows
Editors: Bjorn Engquist, M. Luskin, Andrew Majda

Volume 13: Numerical Algorithms for Parallel Computer Architectures
Editor: Martin H. Schultz

Volume 14: Mathematical Aspects of Scientific Software
Editor: J.R. Rice

Volume 15: Mathematical Frontiers in Computational Chemical Physics
Editor: D. Truhlar

Volume 16: Mathematics in Industrial Problems
by Avner Friedman

Volume 17: Applications of Combinatorics and Graph Theory to the Biological and Social Sciences
Editor: Fred Roberts

Volume 18: *q*-Series and Partitions
Editor: Dennis Stanton

Volume 19: Invariant Theory and Tableaux
Editor: Dennis Stanton

Volume 20: Coding Theory and Design Theory Part I: Coding Theory
Editor: Dijen Ray-Chaudhuri

Volume 21: Coding Theory and Design Theory Part II: Design Theory
Editor: Dijen Ray-Chaudhuri

Volume 22: Signal Processing: Part I - Signal Processing Theory
Editors: L. Auslander, F.A. Grünbaum, W. Helton, T. Kailath, P. Khargonekar and S. Mitter

Volume 23: Signal Processing: Part II - Control Theory and Applications of Signal Processing
Editors: L. Auslander, F.A. Grünbaum, W. Helton, T. Kailath, P. Khargonekar and S. Mitter

Volume 24: Mathematics in Industrial Problems, Part 2
by Avner Friedman

Volume 25: Solitons in Physics, Mathematics, and Nonlinear Optics
Editors: Peter J. Olver and David H. Sattinger

Volume 26: Two Phase Flows and Waves
Editors: Daniel D. Joseph and David G. Schaeffer

Forthcoming Volumes:

1988-1989: *Nonlinear Waves*

Nonlinear Evolution Equations that Change Type

Computer Aided Proofs in Analysis

Multidimensional Hyperbolic Problems and Computations (2 Volumes)

Microlocal Analysis and Nonlinear Waves

Summer Program 1989: *Robustness, Diagnostics, Computing and Graphics in Statistics*

Robustness, Diagnostics in Statistics (2 Volumes)

Computing and Graphics in Statistics

1989-1990: *Dynamical Systems and Their Applications*

An Introduction to Dynamical Systems

Patterns and Dynamics in Reactive Media

Dynamical Issues in Combustion Theory

FOREWORD

This IMA Volume in Mathematics and its Applications

TWO PHASE FLOWS AND WAVES

is based on the proceedings of a workshop which was an integral part of the 1988-89 IMA program on NONLINEAR WAVES. The workshop focussed on the development of waves in flowing composites. We thank the Coordinating Committee: James Glimm, Daniel Joseph, Barbara Keyfitz, Andrew Majda, Alan Newell, Peter Olver, David Sattinger and David Schaeffer for planning and implementing the stimulating year-long program. We especially thank the Workshop Organizers, Daniel D. Joseph and David G. Schaeffer for their efforts in bringing together many of the major figures in those research fields in which modelling of granular flows and suspensions is used.

Avner Friedman

Willard Miller, Jr.

PREFACE

This Workshop, held from January 3-10, 1989 at IMA, focused on the properties of materials which consist of many small solid particles or grains. Let us distinguish the terms *granular material* and *suspension*. In the former, the material consists exclusively of solid particles interacting through direct contact with one another, either sustained frictional contacts in the case of slow shearing or collisions in the case of rapid shearing. In suspensions, also called two phase flow, the grains interact with one another primarily through the influence of a viscous fluid which occupies the interstitial space and participates in the flow. (As shown by the lecture of I. Vardoulakis (not included in this volume), the distinction between these two idealized cases is not always clear.) Both kinds of materials exhibit properties analogous to a fluid (they can flow freely and undergo large deformations) and analogous to a solid (e.g., a granular material can sustain some shearing stress at rest). Both kinds of flows have important industrial and geological applications; the fluidized beds used in many industrial processes deserve special emphasis in this connection.

These materials raise a number of new scientific questions. For the most part, the speakers at the Workshop used a continuum theory in their approaches to these problems. (However, the lecture of O. Walton (not included in this volume) considered a direct, particle-by-particle numerical simulation.) The majority of the issues studied revolve around the constitutive behavior of such materials. We are still far from a rigorous, complete derivation of continuum constitutive behavior. Several authors addressed this challenging problem. Others analyzed theoretically or solved numerically the partial differential equations which result when an *ad hoc* constitutive law is assumed; such a law may be based on experiment or on theoretical considerations. Other papers reported on experimental phenomena exhibited by such materials. Still others considered the application to fluidized beds.

The study of such materials is still at an early phase of its development, and the unresolved questions greatly outnumber the answered ones. These problems have been worked on by researchers from several different fields (engineering, physics, mathematics). In such an underdeveloped area it seems important to consider the perspectives of all these different fields. The purpose of the Workshop was to assist in combining these viewpoints by putting researchers in different fields in close contact with one another. If the lively discussions at the Workshop were a reliable indicator, the Workshop was very productive towards this end.

The contents of specific papers are as follows. First, regarding two phase flow, the papers of Drew, Arnold, and Lakey, of Jenkins and McTigue, and of Passman studied the fundamental problems of deriving constitutive behavior theoretically. Prosperetti and Satrape analyze the stability of various models for two phase flow, while Wallis analyzes one such model in detail. The papers of Gibilaro, Foscolo, and di Felice and of Singh and Joseph concern the important application of fluidized beds. Regarding granular flow, Baxter, Behringer, Fagert, and Johnson report on experiments. In separate papers, Collins and Schaeffer study mathematical properties of equations describing granular flow with an assumed constitutive law. Pitman presents the results of the numerical solution of such equations. (Remark: It is noteworthy that no paper attempts to derive from micromechanics the constitutive behavior of a granular material in the friction-dominated slow flow regime. This absence is a reflection of the difficulty of the problem, not simply an oversight.)

There remains only the pleasant duty of thanking the lecturers for their stimulating contributions and the IMA staff for its courteous, efficient support of the Workshop.

Daniel D. Joseph
David G. Schaeffer

CONTENTS

CONTENTS

PATTERN FORMATION AND TIME-DEPENDENCE IN FLOWING SAND

G.W. BAXTER†, R.P. BEHRINGER†, T. FAGERT‡, AND G.A. JOHNSON‡

Abstract. We describe three experiments characterizing the dynamics of sand flowing from a hopper. In the first experiment, we have measured time-dependence in the normal stress caused by sand flowing from a hopper in order to test recent predictions by Pitman and Schaeffer. In the second experiment, we have used fast X-ray transmission imaging techniques to follow the evolution of patterns and the propagation of fronts associated with the patterns. The patterns are seen in rough faceted sand but not in smooth sand of the same size. In mixtures of rough and smooth sand, the amplitude of the patterns vanishes smoothly as X, the mass fraction of rough sand, falls below 0.22. In the third experiment we looked for the change between mass and funnel flow. In funnel flow, the material is stagnant near the hopper walls, and the flow is typically complex; in mass flow there is smooth "laminar" flow throughout the hopper. Contrary to expectations, funnel flow was observed over a wide range of hopper opening angles θ, spanning $19^0 < \theta < 80^0$. Finally, we describe cellular automata models for the flow of sand which reproduce a number of the features seen in the experiments.

1. INTRODUCTION

The flow of granular material is an important technological problem which still presents numerous puzzles[1]. One aspect of flowing granular material which has been relatively unstudied is that of time-dependence, and it is this issue which we address here. The present experiments involve flow in a hopper. We have chosen this kind of flow for study because it is both technologically important and a relatively easy geometry in which to carry out well controlled experiments. We will describe the results from three types of experiments, and in the remainder of the Introduction, we will provide the motivation for each of these.

The first experiment, which is described in Section II, consisted of measurements of the time-dependent normal stress exerted on the wall of a conical hopper as it was emptied of sand. The impetus for these studies was the analysis by Schaeffer and Pitman[2,3] which indicated that instabilities occurring on the failure side of the yield surface should occur on two natural time scales. We expected that flow characterized by these time-scales would occur during the flow of sand from a hopper, and that measurements of some time-varying quantity would show spectral features dominated by these natural times.

One motivation for the second set of experiments came from our studies of the time-dependent stresses in the first experiment. During the course of a run, well defined circular disturbances evolved on the upper free surfaces of the sand. These disturbances implied the existence of internal structure. A convenient tool for determining this structure is X-rays[4–6]. However, the conical geometry of the

---

†Department of Physics and Center for Nonlinear Studies, Duke University. This work supported by the National Science Foundation under Low Temperature Physics grant No. DMR-8714862 and DMS-8804592. RPB would like to acknowledge the hospitality of the Nonlinear Dynamics Group of the University of Texas during the writing of this paper.

‡Department of Radiology, Duke University Medical Center.

first experiment was not suitable for this technique. A thin flat hopper, described in Section III, was used for these experiments. (See also reference 15).

The flat hopper used in the X-ray studies was made of plexiglass so that it was also transparent to visible light. With this apparatus, it was easy to see that flow typically occurred in the central part of the hopper, with stagnant regions toward the outer edges. This type of flow is referred to as funnel flow. It is expected, on the basis of the Jenike radial solution[7] that, unlike funnel flow, steady flow completely filling the hopper should occur when the hopper angle is small enough. This steady flow is referred to as mass flow. The aim of the third experiment was to characterize the expected transition between funnel flow and mass flow as the hopper angle θ was varied. However, contrary to our expectations, we did not observe a transition between the two types of flow, even though we explored a broad range of $\theta : 19^0 \leq \theta \leq 80^0$. As we explain in Section IV, time-dependent states characterized by an interface between stagnant and flowing sand were always observed.

In Section V we consider cellular automata models for granular flow. These models consist of simple iterative rules which are much simpler than the full ‘microscopic" equations, i.e. Newton's equations for a collection of particles. By sacrificing some of the details of the full equations of motion, we can follow many more particles. Thus, cellular automata have the potential to provide otherwise inaccessible details of the microscopic processes. This information can supplement studies of the full equations of motion and the phenomenological continuum models such as critical state soil mechanics.

The final section of this work, Section VI, contains concluding remarks.

II. NATURAL TIME-SCALES FOR HOPPER FLOW

A. Predictions. Pitman and Schaeffer[2,3] have shown that instability in granular flows should occur on two different time-scales. These are a slow time scale

$$t_s = \beta L/U,$$

and a fast time scale

$$t_f = U/g.$$

Here, U is a characteristic speed for the flow, and L is a length which is nominally determined by the geometry of the experiment. Also, g is the acceleration of gravity, and β parameterizes the yield surface of the granular material. The point of the present experiments was to determine if, at least, the slow time scale could be observed in a simple experiment.

Some additional details of the theory are pertinent here. The analysis assumes that the flow of granular materials can be described within a continuum plasticity theory. For instance, given two-dimensional flow, deformation will occur at a density ρ, when the shearing stress $\tau = (\sigma_1 - \sigma_2)/2$ is sufficiently large with respect to the hydrostatic stress $\sigma = (\sigma_1 + \sigma_2)/2$ that the yield surface is encountered. (Here,

σ_1 and σ_2 are the principal stresses, i.e. the eigenvalues of the stress tensor.) An approximate yield surface is given by

$$\tau(\sigma,\rho) = (\tau_0/p)\{p^2 \sin(\delta) - C[(\sigma/\sigma_0) - p]^2\},$$

where $p = (\rho/\rho_0)^{1/\beta}, \beta$ is a material constant, as are the other constants, C, τ_0, σ_0, and δ, the angle of internal friction. For a material like sand, β is about 10^{-2}.

The time scales identified by continuum theory must be augmented by a microscopic time scale, t_m, given by

$$t_m = d/U,$$

where d is a characteristic dimension for a single grain of sand.

B. Time-dependent stresses in a cylindrical hopper-experimental details. In order to test for the time-scales of Pitman and Schaeffer, we carried out measurements of the time-dependent stress in a conical hopper, which is sketched in Figure 1. The hopper was 0.5 m across at the top and terminated at the bottom in a 2.5 cm opening. The full cone angle, θ, was 60^0.

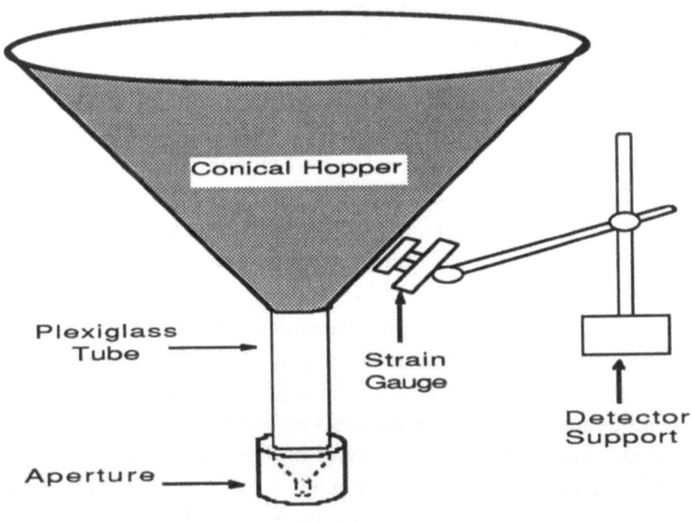

Figure 1. Schematic of the conical hopper used to measure the time-dependent stress. A cylindrical extension beneath the cone holds a variable aperture that sets the flow rate. The rigid arm, which holds one of the capacitor plates, is also shown.

Using the dimensions appropriate to our experiments, $L \approx 10cm, d \approx 0.06cm$, an estimate for the speed of $U \approx 0.1cm/s$, and $\beta \approx 10^{-2}$, we obtain $t_s \approx 0.1s, t_f \approx$

$10^{-4}s$, and $t_m \approx 0.6s$. Since the experiments actually yield data in terms of frequencies, it is convenient to define characteristic frequencies

$$f_s = 1/t_s,$$
$$f_f = 1/t_f,$$

and

$$f_m = 1/t_m,$$

so that for our experiments, $f_s \approx 10Hz, f_f \approx 10^4 Hz$, and $f_m \approx 2Hz$. Assuming that L and β are constant, we would expect that f_s would vary linearly with U. The frequencies f_s and f_m are well enough separated that we expect to be able to distinguish them. However, the fast frequency f_f is too large to be resolved with the current experimental equipment.

Some additional details of the experimental apparatus are appropriate here. In order to provide a controlled flow rate, we terminated the hopper by a 2.5 cm diameter tube, which in turn ended in a tapered aperture. To a good approximation, flow in the tube occurred as a solid body, so that any perturbations caused by the output aperture had a negligible effect on the flow of interest. A range of interchangeable aperture pieces allowed us to vary the mass flow rate, M, over the range $2gram/s \leq M \leq 32gram/s$. For the slowest rates, we used the smallest aperture which would remain unblocked during the course of a run. The upper bound on M was chosen so that at the end of a run, the sand level was still well above the stress transducer. The flow rate as a function of aperture diameter D is given in Figure 2. As expected from previous work[8], the flow rate was constant[9], independent of the height of the sand, until the sand level was almost down to the bottom of the cone. The data of Figure 2 are well described by $M = AD^\alpha$, with $A = 309$, and $\alpha = 2.96$.

The stress transducer measured time-dependent normal stress exerted by the sand on the hopper wall. The transducer design exploited the fact that the hopper was constructed of a thin resilient sheet of brass (thickness 0.5 mm) supported on a rigid frame of aluminum rings. The brass sheet deformed very slightly in response to the normal stress from the sand. The amount of deformation was always less than 0.005 cm, but by using a capacitive technique, we could measure the deformation, and hence the stress, with considerable precision. At a location about one third the way up from the cone opening, on the outside of the brass sheet, we glued a thin (0.005 cm thick) copper foil, approximately 2 cm on a side, and insulated from the brass wall of the cone by mylar. This foil formed one plate of a parallel plate capacitor. The other capacitor plate was held close to the copper foil by a rigid damped bar. Changes in the stress were then transduced into changes in this capacitor, which we refer to as C_v (variable).

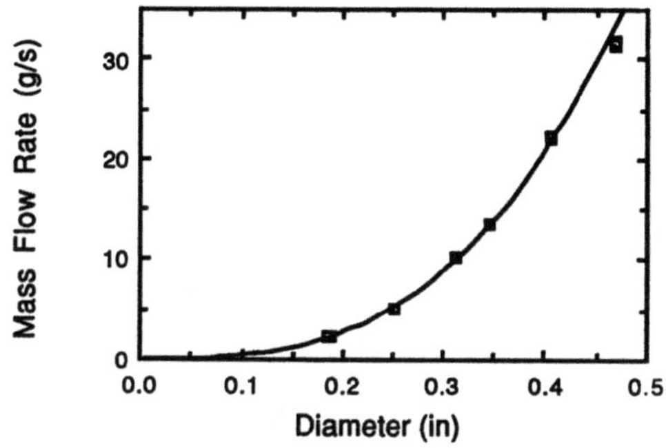

Figure 2. The mass flow rate M as a function of aperture diameter D for a rough faceted sand of diameter $0.06 \pm 0.02 cm$.

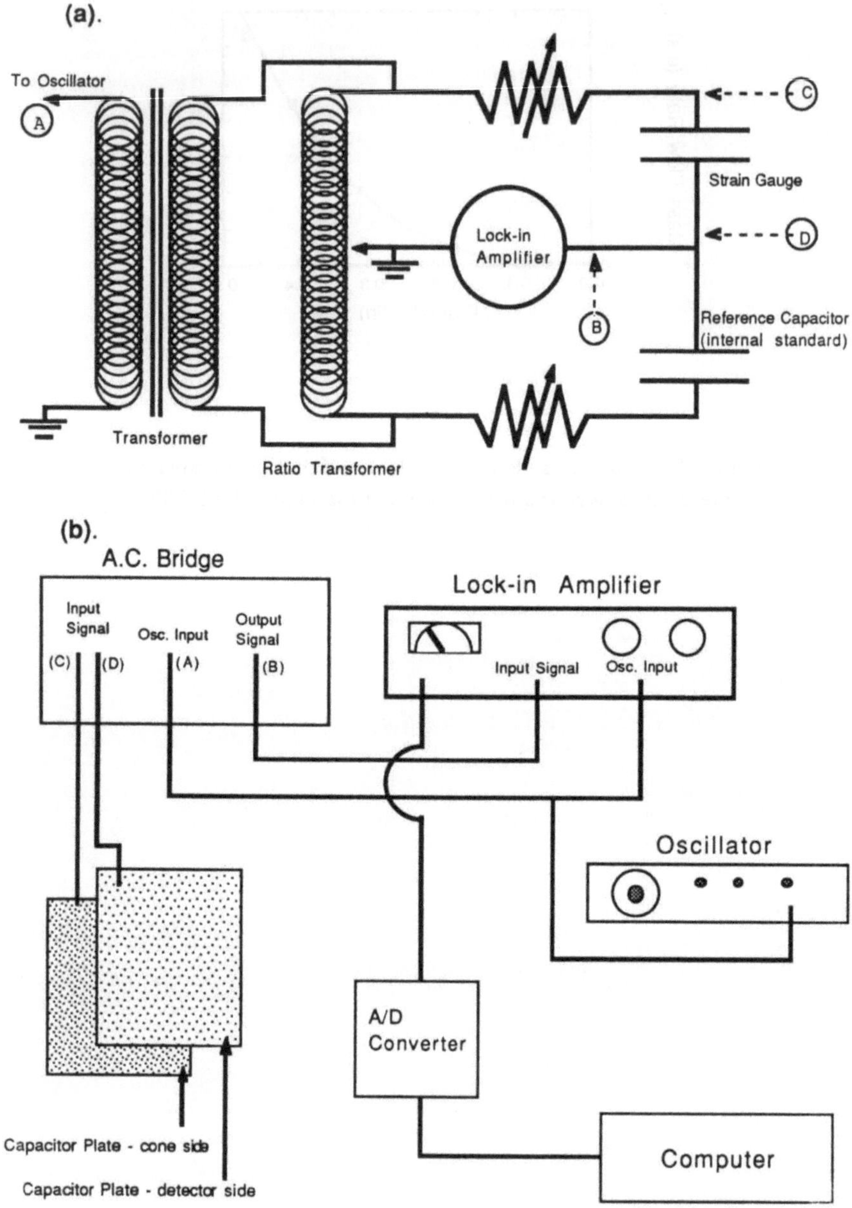

Figure 3. a. Bridge circuit to measure C_v. b. Schematic of the full electronics for obtaining time-series measurements of C_v.

The capacitance C_v could be precisely measured with a bridge circuit sketched in Figure 3a. Either commercial (General Radio) or homemade bridges were used with similar results. The typical resolution of C_v was about one part in 10^5 with a time constant of 1 ms or 4 ms.

During the course of a run, time-series data were acquired using the arrangement of Figure 3b. The key feature of this arrangement was that time-variations of C_v led to corresponding variations in the out-of-balance signal of the lockin amplifier. The out-of-balance signal of the lockin was then recorded at regular time intervals Δt by a microcomputer equipped with analogue-to-digital converters. Typically, we used $0.001s < \Delta t < 0.01s$, where $\Delta t = 0.001s$ was at the limit of the computer's data recording ability.

C. Time dependent stresses in a cylindrical hopper - results. Two examples of the resulting time series are given in figures 4a and 4c. The first of these gives the stress (in arbitrary units) versus time for a typical run; part c of the figure shows the signal when the hopper is full of sand, but there is no flow. Since both Figures 4a and 4c were obtained at the same sensitivity, a comparison gives an indication of the signal to noise ratio.

In order to analyze the experimental results, several steps are necessary. From the time-series data, typified by Figure 4a, we have removed the slowly varying part by fitting the complete run to a low order polynomial. We attribute this slowly varying part to the gradual emptying of the hopper. The remaining more rapidly varying part, which we view as the interesting time-dependence, is typified by Figure 4b. Specifically, to obtain figure 4b, we have subtracted away the slowly varying low-order polynomial fit which is shown as the solid line in Figure 4a.

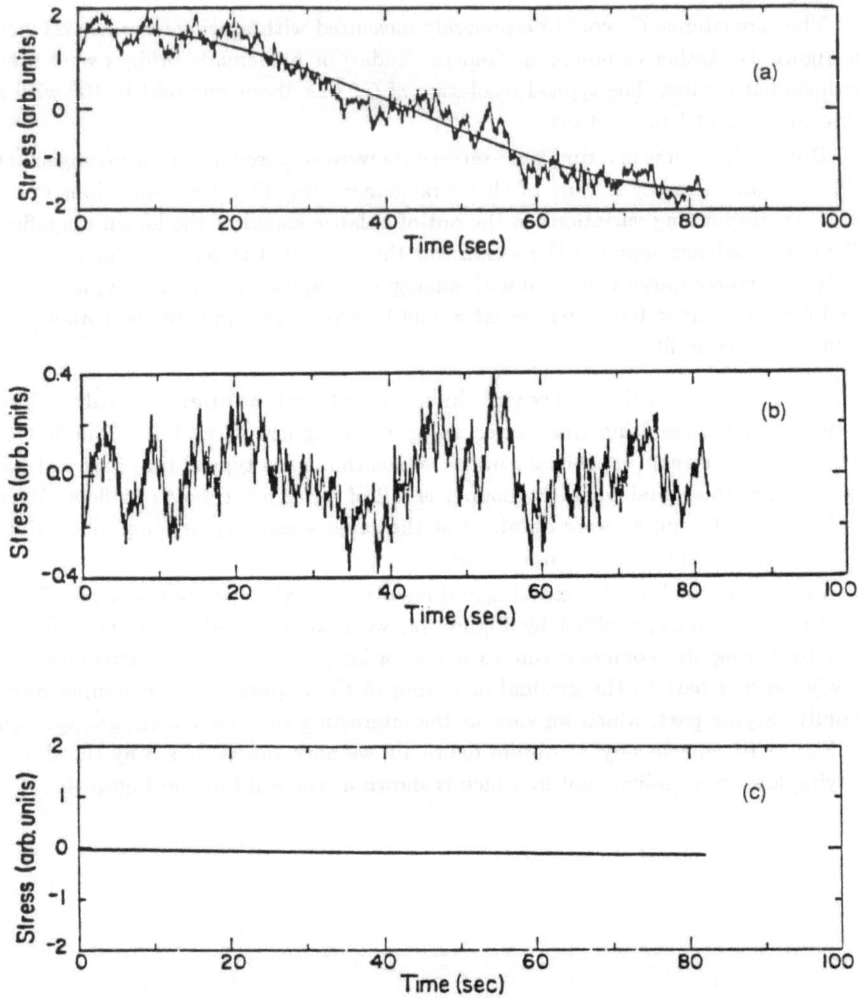

Figure 4. a. Variation versus time, t, of the normal stress in arbitrary units, as measured by the circuits of Figure 3. The solid line is a third order (in t) polynomial fit to the slowly varying background. b. The rapidly varying portion of the time series of part a obtained by subtracting the solid curve from the data. c. The signal obtained when the hopper was filled with sand, but there was no flow.

In order to characterize the time-dependence we have computed the power spectrum, $P(f)$, defined in terms of the time varying signal $S(t)$ as

$$P(f) = T^{-1} \left| \int S(t) \exp(-2\pi i f t) dt \right|^2,$$

where the integration is carried out over the observation time $0 \leq t \leq T$, and the Fourier integral was computed with a standard FTT algorithm[10]. For example, Figure 5a was the spectrum obtained from Figure 4b. Note that for frequencies well below the Nyquist frequency[10], the power spectrum of the instrumental noise falls several orders of magnitude below the spectrum due to the flow. An example of the instrumental noise power spectrum, obtained from the time-series of Figure 4c, is given in Figure 5b. (Note that in figure 5, the logarithms of P and f have been used.)

Figure 5a is striking by its absence of any sharp features. In particular, no *sharp* feature is located near the slow frequency f_s associated with the instability analysis of Pitman and Schaeffer[2,3]. However, in Figure 5a, there is, near the expected f_s, a change in the slope of $\log(P)$ vs. $\log(f)$ from roughly -2 to roughly -4. This slope change is always seen. Therefore, it is interesting to compute f_k, the frequency of the slope change, or "knee frequency" as a function of the characteristic speed U. Here, we take U to be the speed of the sand just as it leaves the bottom of the cone and enters the top of the cylindrical tube of Figure 1. (U is easily determined from the density of the sand, the mass flow rate, and some geometric parameters.)

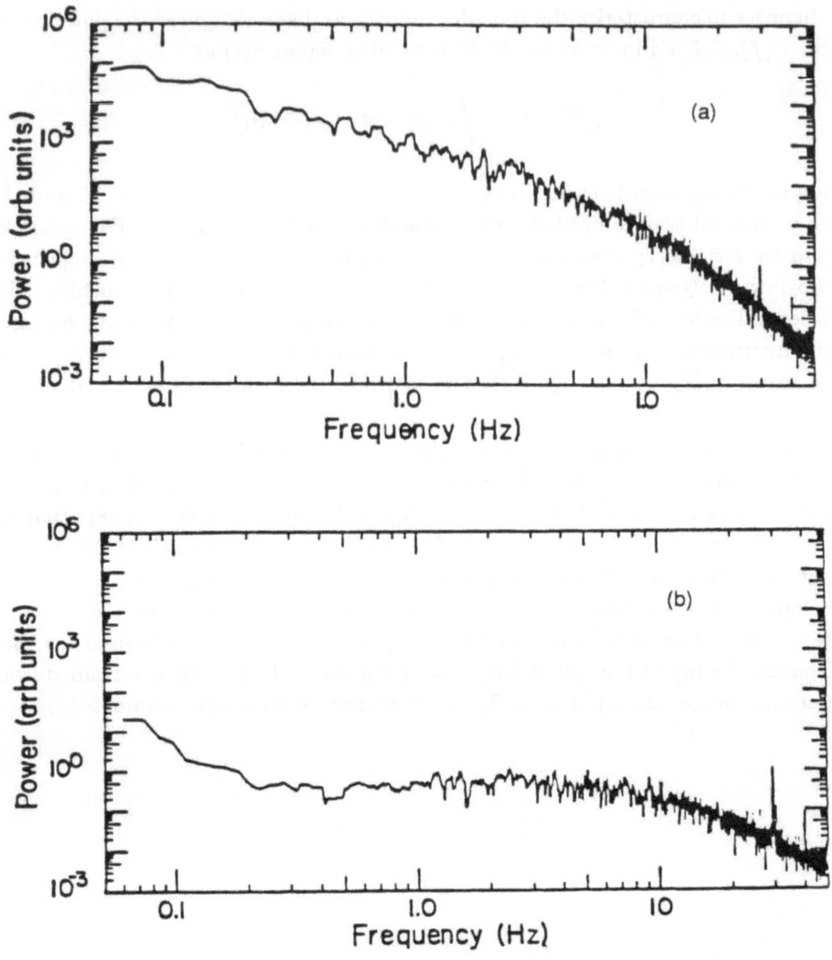

Figure 5. a. Spectrum obtained from Figure 4b,which is the "fast" time-dependence seen in the normal stress at the sidewall of the hopper. b. Spectrum for the instrumental noise spectrum, in this case, the spectrum obtained from figure 4c.

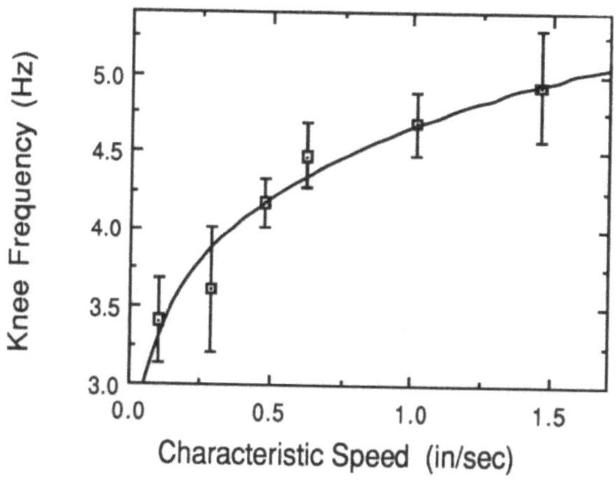

Figure 6. The knee frequency, f_k, versus the characteristic speed, U, at the exit of the cone. The solid line is a least-squares fit with parameters given in the text.

We show in Figure 6 results for f_k as a function of the characteristic speed U. Contrary to our expectations, we find

$$f_k(U) = AU^{\alpha},$$

with $A = 4.66$ and $\alpha = 0.15$. This is quite different from our expectations for $f_k(U)$, namely, $f_k \propto U^{1.0}$.

The experiments raise a number of questions regarding the roles of microscopic processes, the propagation of time-dependent signals through the medium, and the evolution of time varying patterns, possibly through nonlinear processes. Also, one must question how accurately the experiments yield the true normal stress of the walls: Does the flexible nature of the walls everywhere limit our ability to make a local stress measurement? Is the frequency response of the transducer flat to high enough frequencies? (A rough determination of the frequency response indicates that the answer is probably yes.) By removing the slowly varying background, have we inappropriately changed the character of the spectra?

In order to rule out any effects due to instrumentation, we are building an apparatus which will only measure local stresses with a well defined frequency response. A differential stress technique will be used which is insensitive to the stress drop as the height of the sand decreases. Also, in the following section we describe measurements which, at least partially, address the issue of spatial dynamics.

III. DIGITAL SUBTRACTION FLUOROSCOPY OF SAND FLOW IN A FLAT HOPPER

A. Experimental apparatus. We begin this section by providing a number of experimental details of the apparatus used to obtain time-dependent images of the density of sand undergoing flow in a thin flat hopper.[15] The hopper geometry, indicated in Figure 7, was chosen for its intrinsic practical interest and for its amenability to X-ray transmission measurements. Parallel X-rays traversed the thin dimension of the hopper (Figure 7). The intensity of the X-rays was then detected by a plumbicon (an X-ray version of a vidicon) and the image was recorded digitally. The complete system was a clinical X-ray machine consisting of an ADAC digital subtraction system coupled to a Phillips fluoroscopic imaging chain.

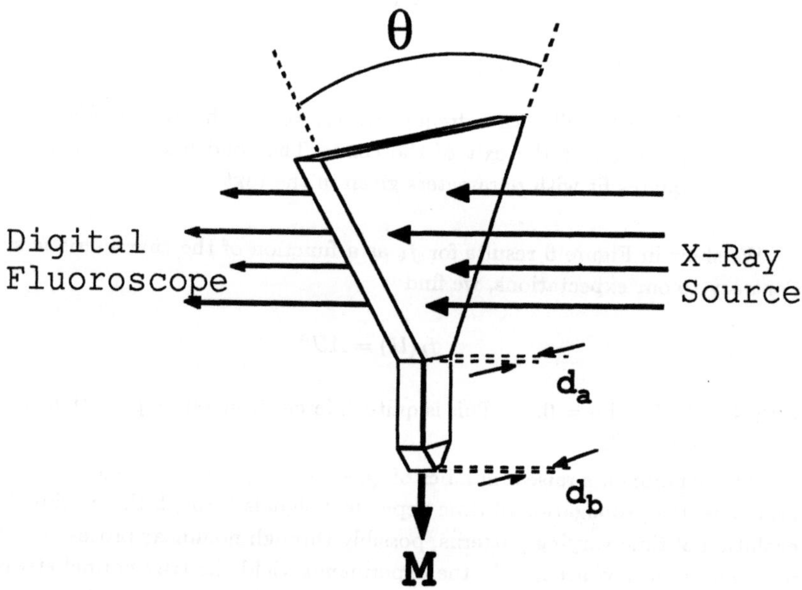

Figure 7. Schematic of the flat hopper for X-ray digital subtraction fluoroscopy.

As with the conical hopper, the flow rate was set by means controlled by means of an adjustable aperture located on an extension beneath the outlet of the hopper. The flow rate was set by the size of the aperture d_b, so long as d_b was smaller than d_a. When d_b was greater than d_a, the sand went into free fall at the outlet of the hopper, i.e. at d_a.

This device provides three significant improvements over past X-ray imaging of hopper flows. First, the images were obtained at a fast sampling rate, 10 images/s. In past X-ray studies, the flow was typically stopped to obtain each exposure, so that the details of the pattern evolution were not readily available. Also, the experimenters had to assume that stopping or starting the flow was a non-perturbational process. Second, the transmitted X-ray intensities were digitized with a spatial resolution of 0.05 cm, a bit smaller than the mean diameter of a sand grain. Therefore, quantitative data were available on small spatial scales. Third, as a consequence of point two, the images could be significantly enhanced by digitally subtracting the background, i.e. by subtracting from each image of the flow, an image taken with the hopper filled with sand, but without any flow.

Several physical dimensions are pertinent. The hopper was 1.27 cm thick and 30 cm high at the highest point. The opening angle θ, could be conveniently varied over $19^0 \leq \theta \leq 90^0$. The field of view was 25 cm in diameter with a resolution of 512×512 pixels. Two types of sand were used in the experiments. The first was a rough faceted sand of diameter $0.06cm \pm 0.02cm$. The other was a nearly spherical sand (Ottawa sand) of diameter 0.07 cm with a somewhat smaller variation about the mean diameter.

B. Experimental results for pattern formation. We consider first the flow of a rough faceted sand of size $0.06cm$. Figure 8 shows a subset of the images obtained every $0.1s$ with $\theta = 66.8^0$ and $M/M_f = 1.0$ ($M_f = 15.2g/s$ is the free-fall value of M). This figure shows very clearly the presence of propagating modes. (Here, and throughout this paper, darker regions of the image are associated with lower densities compared to brighter regions of the image which are associated with higher densities.) Note that the patterns which form are typically symmetric about a vertical midline. However, a small tilting of the apparatus produced asymmetric patterns.

There are a number of the parameters which might affect the flows. Of particular importance are the opening angle θ of the hopper, and a parameter X, to be described below, which characterizes in some sense the average roughness of the sand particles. Also of some relevance is the mass flow rate M at which material is removed from the hopper.

The fast image rate and the digital subtraction allow us to follow the pattern evolution in detail. During the course of a run, several fronts formed, particularly for small angles. In particular, we show in Figure 9 the position of the first low density front (the upper edge) as seen at the midline of the hopper. Results are given as a function of time for several angles. Results such as those in Figure 9 were reproducible within about 10% , at least for the mean velocity of the front. Note that for small θ the front moves downward, and that for large θ the frontal motion is upward, opposite to the direction of the net mass flux. There is an angle, $\theta \cong 35^0$, for which the fronts are nearly stationary (in the lab frame). We show in Figure 10 example frames for several additional angles.

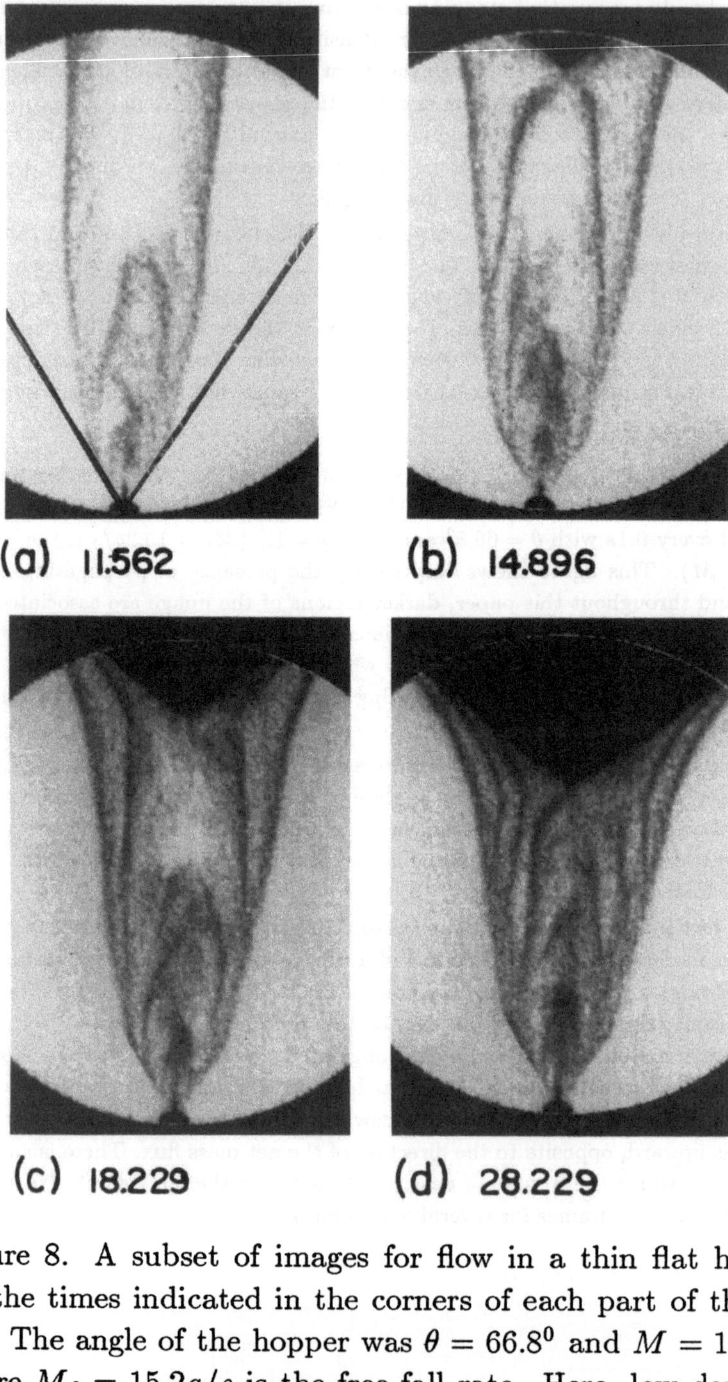

Figure 8. A subset of images for flow in a thin flat hopper for the times indicated in the corners of each part of the figure. The angle of the hopper was $\theta = 66.8^0$ and $M = 1.0M_f$, where $M_f = 15.2g/s$ is the free fall rate. Here, low densities correspond to dark regions.

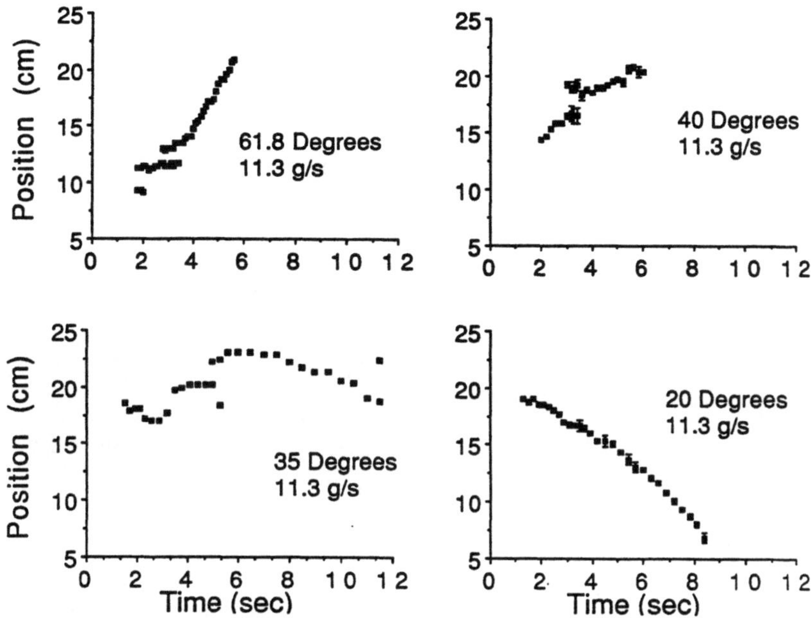

Figure 9. The position versus time for the first observed low density front for each of several angles.

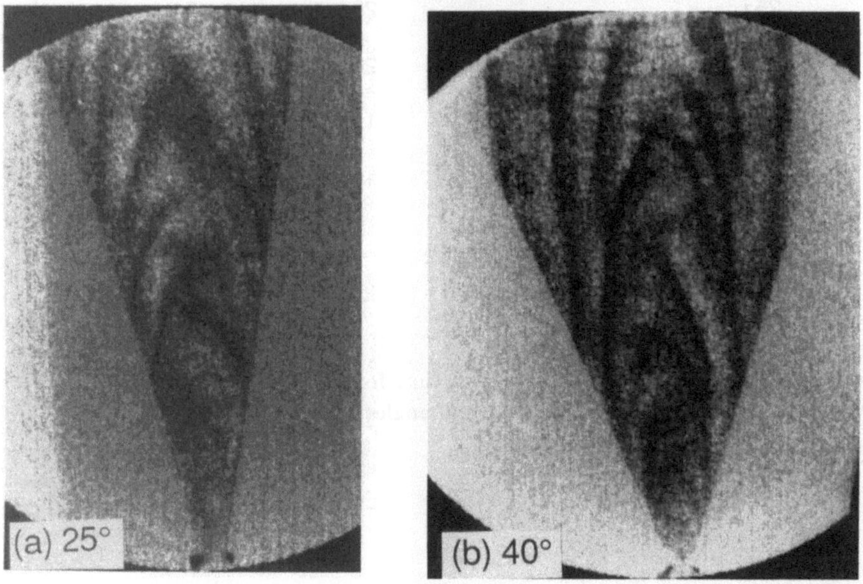

Figure 10. Images for several additional angles. a. Cone angle is 25°. b. Cone angle is 40°.

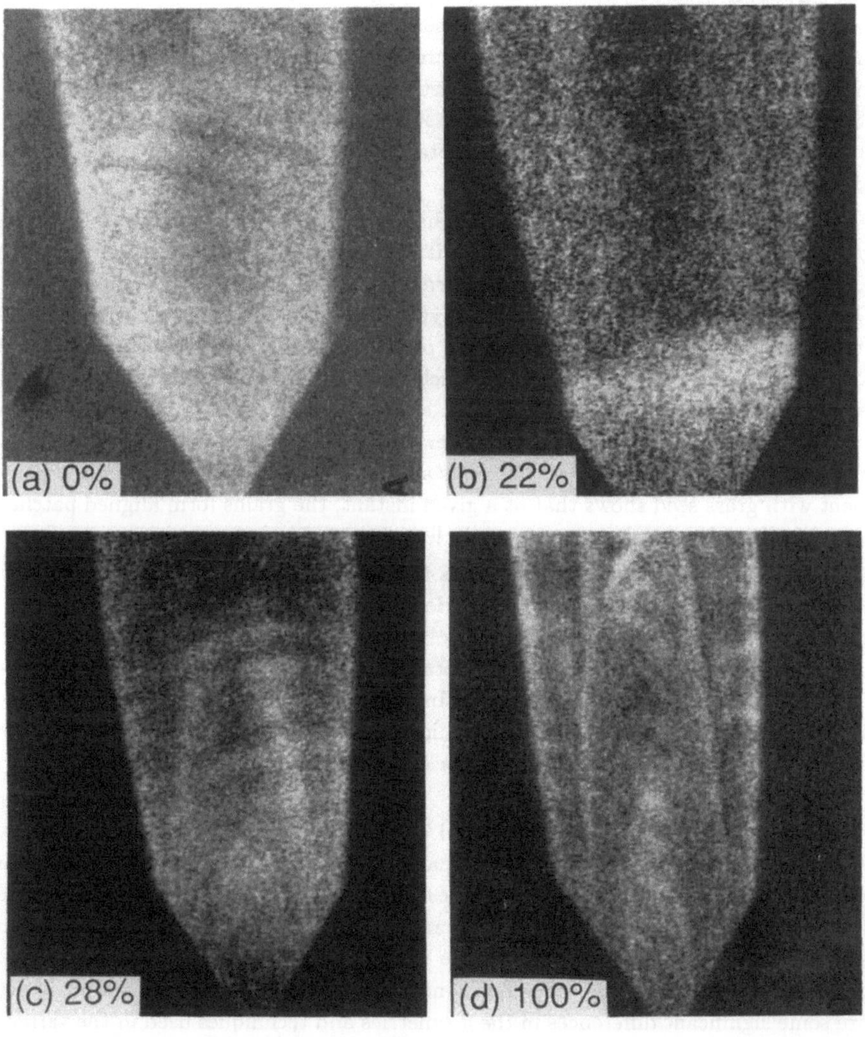

Figure 11. a. A subset of images for flow with smooth sand showing the absence of propagating fronts. b-d. Other parts of the figure show images taken with mixtures of rough and smooth sand, where each mixture is characterized by the mass fraction X of rough sand. Mass fractions are: 0% for a, 22% for b, 28% for c, and 100% for d. The cone angle is 59°.

If the rough sand is replaced by the smooth, nearly spherical Ottawa sand (diameter = 0.7 mm) the propagating waves are not seen at all. We have explored this surprising result by making mixtures of the rough and smooth sands so that each mixture is characterized by X, the mass fraction of rough sand. We do not see any waves below $X \approx 0.22$; but as X increases above 0.22, waves are visible with increasingly strong amplitude. We demonstrate this in Figure 11, and conclude that there may be a forward bifurcation, with X the control parameter. Unfortunately, the data are not yet good enough to obtain quantitative details of the apparent bifurcation.

The experiments show that the roughness of the sand plays a key role in the generation of the waves. In this regard, it is interesting to note that there is relatively little difference in the angles of internal friction of the rough and smooth sand. One possible difference between the two types of sand may be that the rough sand's structure makes the rotation of individual particle an important property. This local rotation field, or Cosserat structure, which has been considered by Vardoulakis[11], is not contained in conventional critical state soil mechanics. Also, a rotational variable is almost surely relevant for material where the individual grains are longer in one direction than the other (i.e. grass seed, rice). For instance, a simple experiment with grass seed shows that at a given instant, the grains form aligned patches over length scales which are many grain lengths.

One aspect of the waves which needs further exploration is their characteristic length scales. For instance, are these lengths set by the container or by some property of the material and the control parameters? This may be particularly important for the frequency scale f_s, obtained from the conical hopper, since this frequency contains the length scale L. In calculating f_s, we assumed that L was set by the container geometry, and remained fixed as U was varied. However, the flow may contain its own natural length scale which could be more relevant than the container size.

Before closing this section, we should compare our results to previous work[4–6]. A number of X-ray studies have been made of hopper flow. One of the points on which our work differs with that of Drescher et al.[4], for example, is that we see flow with symmetry about the vertical midplane, whereas the earlier work often shows shear bands which arch from one side to the other in an asymmetric way. A good explanation of this difference is not immediately available. However, there are some significant differences in the geometries and techniques used in the various experiments.

IV. STAGNANT/FLOWING INTERFACE

One characteristic of the flow in a hopper is the formation of an interface between moving sand in the center of the hopper and a stagnant region next to the walls. This interface can be seen by eye in the flat hopper described in the previous section. It is also visible in X-ray images because in the vicinity of the interface, there is a band of low density.

The Jenike radial solution[7] predicts that for small hopper angles there will be no interface (mass flow) whereas for large hopper angles there will be an interface (funnel flow). We expected that a transition between mass and funnel flow should be observable as θ was varied.

The last set of measurements described in this paper involved a determination of the position of this interface relative to the opening of the hopper as a function of θ. As θ was decreased, the position of the interface (d_s in Figure 12a) moved upwards, away from the vertex of the hopper. In order to determine the steady state position of the interface at small angles where d_s became large, we constructed a tall hopper-two meters high - with the same thickness, $1.27cm$, as that used for the X-ray studies. This hopper was $60cm$ wide, and constructed of plexiglass. With it, we could visually determine the position of the interface for angles as small as 19°.

In Figure 12b, we have displayed, as a function of θ, data for the distance d_s (indicated in Figure 12a) between the apex of the hopper and the stagnant/flowing interface. Data have been obtained for the small adjustable hopper used for X-ray experiments and the two-meter hopper. In each case we used the rough faceted sand described previously. The results of Figure 12b are presented in the form d_s^{-1} vs. θ, since we anticipated that for some θ within our experimental range, d_s^{-1} would vanish. This expectation was not borne out. Rather, the data are well described by $d_s^{-1} = B(\theta/\theta_0)^\beta$, with $B = 1.00cm^{-1}, \beta = 2.252$, and $\theta_0 = 121.0^0$ (a convenient reference angle). This power law behavior is seen very clearly if logarithmic scales are used as in Figure 12c. In these experiments, there is always a static region. However, the location is so high up the hopper for small θ, that in most practical situations (where a vertical bin feeds the hopper) d_s would be larger than the hopper length. Mass flow will result, but as a "finite size" effect. Note too, that even for small θ, throughout the small hopper the flow is time dependent. Thus, it does not appear that the radial solution is realized in these experiments.

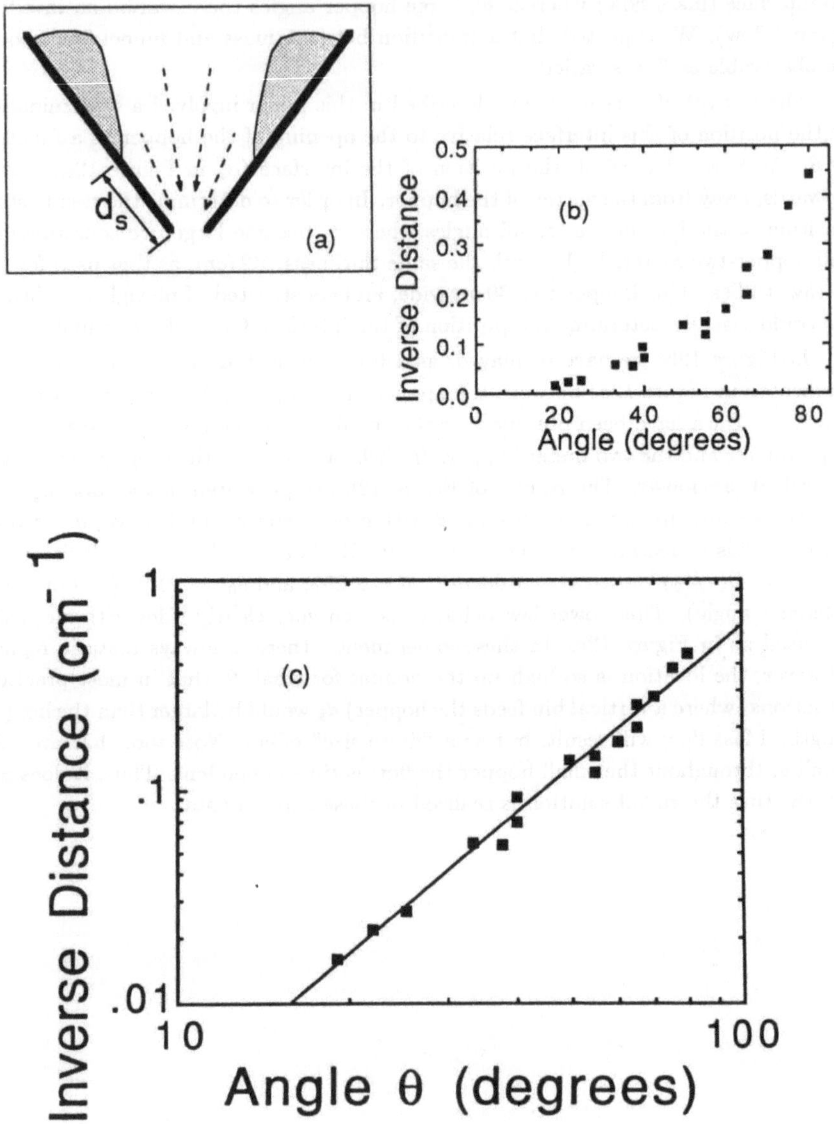

Figure 12. a. Schematic of the hopper showing the distance d_s, the distance between the apex of the hopper and the stagnant/flowing interface. b and c. $(d_s)^{-1}$ versus θ on linear and logarithmic scales. The solid line is the power law fit to the data.

V. CELLULAR AUTOMATA MODELS

A potentially useful but virtually unexplored tool for understanding granular flows is to be found in cellular automata models[12–14]. These consist of simple iterative rules for a set of "particles" where the rules are far simpler than the complete Newtonian dynamics for real particles. However, the rules are chosen to contain fundamental features of the flow.

Automata have both advantages and disadvantages relative to other techniques such as continuum theories and full numerical integration of Newton's laws. Among the advantages are the large number of particles which an automaton may contain (comparable to the number of particles in a physical experiment with sand), the relative ease with which it can be implemented, and the small amount of computer time needed to describe a flow. These models are not complicated by some of the numerical problems which occur in the continuum models. They can provide insight into the key physical features of a flow, and that insight is available on the microscopic length and time scales as well as the longer scales appropriate to continuum mechanics. The outstanding disadvantage is that automata models are heuristic at best and certainly not exact. An interesting issue is whether automata provide worse, better, or simply different descriptions than continuum models where the physical modelling and approximations occur in the constitutive relations. Here, we present results for two automata, although we do not attempt a detailed justification.

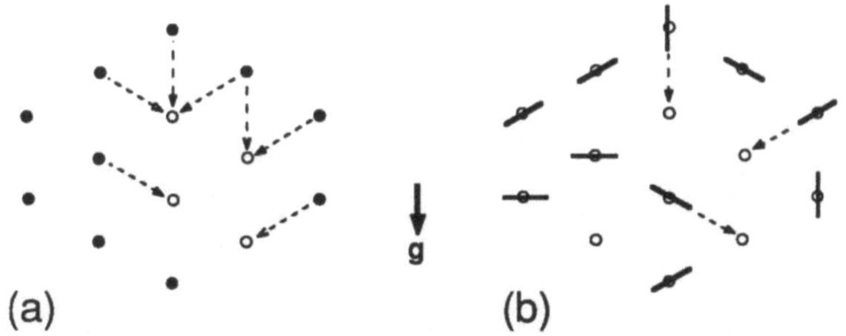

Figure 13. a. Schematic of possible flows of structureless particles on a hexagonal lattice. b. Similar schematic for particles with an orientational structure.

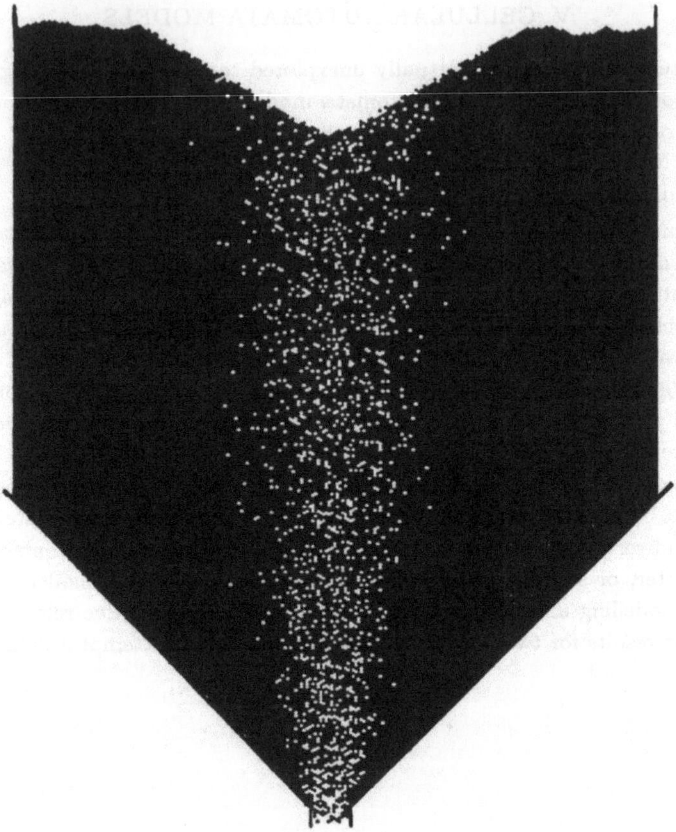

Figure 14. Distribution of structureless particles after 1000 iterations of the automaton of Figure 13a. White regions indicate vacant sites.

Our first automaton applies to structureless "particles" which move from point to point on a regular hexagonal two-dimensional array, as in Figure 13a. We expect that this model most closely approximates the flow of smooth sand. The particles are contained in a hopper geometry, and in response to a gravitational-like force, they move downwards. If they come to the bottom of the opening they are removed. A particle can only move if there is an open site either directly underneath or underneath but slightly to one side (again see Figure 13a). This feature models the relatively dense packing of the physical experiments. When there is more than one option for a particle to fall into an empty site, a choice must be made. Here, we assign a fixed probability of moving in a given direction; then, a random number generator is used to actually determine the outcome of a particular choice. The automaton is updated by examining each row, starting at the bottom and moving upward, and carrying out all appropriate site changes. Each iteration through the complete lattice can be carried out in less than one second. An example showing the distribution of particles after a large number of iterations is given in Figure 14. (This particular example involves a lattice of 65536 sites. Initially 3% of these sites were chosen at random to be holes.) This model shows a reduced-density region through the center of the material, and the V-shaped drop in the upper free surface which is characteristic of physical experiments.

The second automaton which we present here is an attempt to model some of the features seen in the flow of rough sand. Accordingly, we must ascribe to each particle some additional structure. We model this structure by assuming that the "particles" are like little sticks, i.e., that they have an orientation as well as a position (Figure 13b). Such a model is particularly well suited to the flow of grains, such as grass seed or rice. The orientation of a particle relative to its neighbors affects its ability to move. There must be a mutual coupling between the orientational motion and the displacement of the particles. Figure 15 shows the result from the implementation of such a rule after 4096 iterations. (This particular example used a lattice of 262144 sites. Initially, each site was filled with a particle having one of the 4 allowed orientations chosen at random.) This more sophisticated model contains bands of oriented material which may be related to some of the shear bands which form in physical experiments. Of some interest are the spatial and temporal characteristics of the flow. For instance, in Figure 16 we show the spatial variation of the density along the central channel. Part a of the figure shows density versus position, where the density has been computed by averaging over nearest neighbor sites (the upper surface is located near position 410); part b is the spatial power spectrum of the central region of part a (in order to avoid end effects). Note that the spectral density falls off roughly as the inverse of the wave vector. This behavior is unlike a Newtonian fluid where the spatial range is much more limited. It is possible that this broad spectrum of wavelengths may be present in real granular flows, and that it must be considered in constructing an accurate continuum model. In Figure 17, we show a related quantity, the time-dependence of the mass flow out of the bottom of the "hopper". Part a shows the number of "particles" which fall out in an iteration, and part b shows the temporal power spectrum derived from part a. In this figure too, we see a power-law dependence over three decades in frequency in the spectrum. Note, however, that the average flow rate is constant, as in the physical experiments. Additional details of this and related models will be given elsewhere[13].

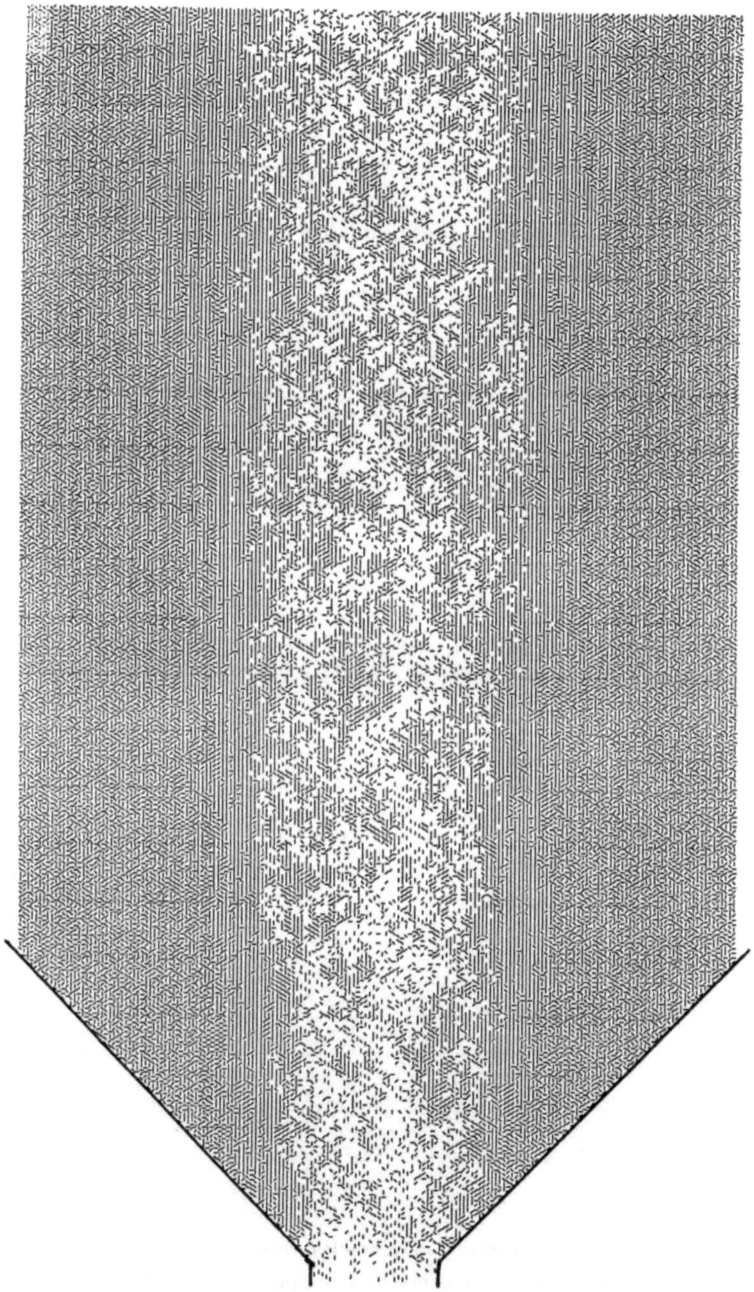

Figure 15. Distribution of particles of the oriented-particle automaton after 4096 iterations. A portion of the lattice is shown. The hopper angle is 60°

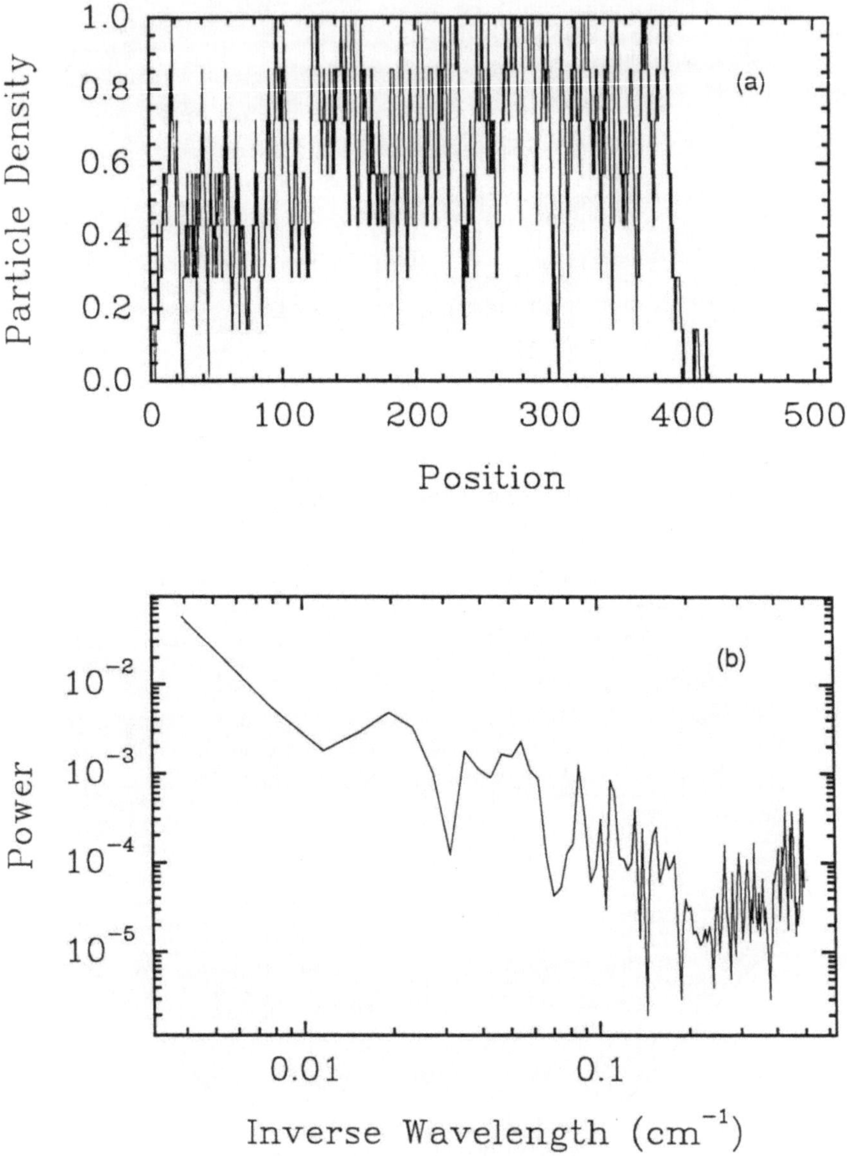

Figure 16. a. Density of particles versus position along the centerline of automata-modeled hopper flow. b. The spatial power spectrum derived from part a of this figure.

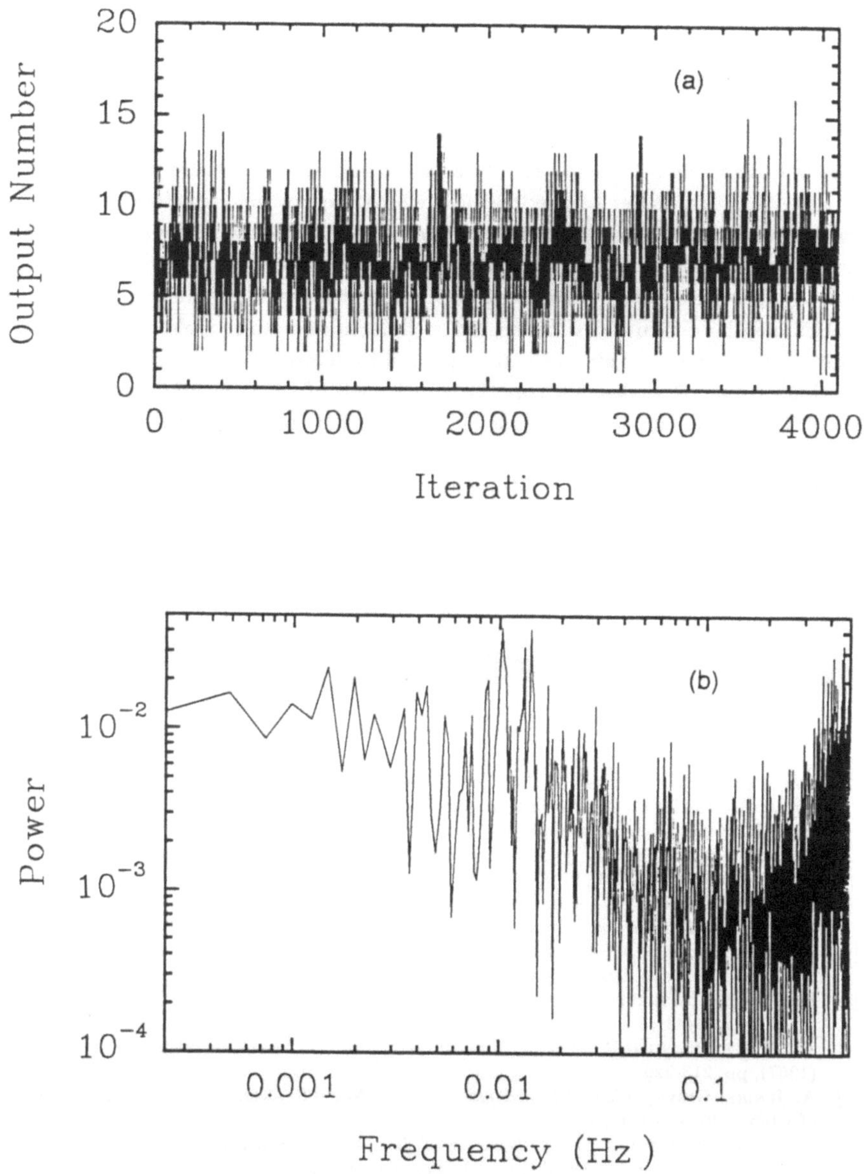

Figure 17. a. Number of particles per iteration falling out of the "hopper" of the structureless-particle automaton. b. Power spectrum obtained from part a of this figure.

VI. CONCLUSIONS

The chief goal of this work has been to provide a better understanding of granular flows by studying their time-varying features. We have found that time-dependence is intrinsic to all the flows which we have studied, even in the regimes where we might have expected steady flow. In the course of these measurements, we have found a number of new features which are not yet explained within the current theoretical framework. Among these are the spectra associated with the normal stress and the propagating modes seen in rough sand but not in smooth sand. A full quantitative treatment of these features within existing continuum theories may be difficult. In this regard, cellular automata models may be useful. Additional experimental studies will be needed to determine the mechanism for the selection of natural scales, such as the wavelength of the propagating modes and the thickness of the associated bands. It is likely that these phenomena are present in three dimensional flows (although we have seen them only in two dimensions), and they may be related to the pulsing phenomena familiar in commercial silos. An understanding of this time-dependence will have both an intellectual and a practical significance.

ACKNOWLEDGEMENTS

We appreciate many helpful conversations with Prof. David Schaeffer.

REFERENCES

[1] For a recent review, see R. Jackson, *Some Mathematical and Physical Aspects of Continuum Models for the Motion of Granular Materials*, in The Theory of Dispersed Multiphase Flow, R. Meyer ed., Academic Press 1983.

[2] D.G. Schaeffer, *Instability in the Evolution Equations Describing Incompressible Granular Flow*, J. Diff. Eq., 66 (1987), pp. 19-50.

[3] E.B. Pitman and D.G. Schaefer, *Stability of Time Dependent Compressible Granular Flow in Two Dimensions*, Comm. Pure Appl Math., 40 (1987), pp. 421-447.

[4] A. Drescher, T.W. Cousens, and P.L. Bransby, *Kinematics of the mass flow of granular material through a plane hopper*, geotechnique 28 (1978), pp. 27-42.

[5] R.L. Michalowski, *Flow of granular material through a plane hopper*, Powder Tech, 39 (1984), pp. 29-40.

[6] J.O. Cutress and R.F. Pulfer, *X-ray investigations of flowing powders*, Powder Tech., 1 (1967), pp. 213-220.

[7] A. Jenike, *Gravity Flow of bulk Solids*, Bulletin No. 108, Utah Eng. Expt, Station, Univ. of Utah, salt Lake City, 1961.

[8] See N. Al-Din and D.J. Gunn, *The Flow of Non-Cohesive Solids through Orifices*, Chemical Engineering Science, 39 (1984), pp. 121-127, and references therein.

[9] M is constant, within a resolution of about 5%, on time-scales which are long compared to t_m, but short compared to the time to obtain a time series or to empty the hopper. Measurements of M were carried out on a time scale of about $10s$ by seeing how long it took to fill small cups. M was then determined by ratio of the mass in the cup to the filling time.

[10] R.K. Otnes and L. Enochson, *Digital Time Series Analysis*, Wiley, N.Y., 1972.

[11] See I. Vardoulakis – elsewhere in these proceedings.

[12] Tommaso Toffoli and Norman Margolus, *Cellular Automata Machines: A New Environment for Modeling*, MIT Press 1987.

[13] G.W. Baxter and R.P. Behringer, to be published.

[14] We are aware of only one other instance in which automata are being applied to granular flows. Work in progress by Drs. Peter Haff, Gary Gutt and Bradley Werner (see also Gary

Gutt, Ph.D. thesis, California Institute of Technology, 1989) uses a considerably different adaptation of cellular automata to granular flow.
[15] G. William Baxter and R.P. Behringer, *Pattern Formation in Flowing Sand*, Phys. Ref. Lett., 62 (1989), p. 2825.

THE MATHEMATICAL STRUCTURE OF THE EQUATIONS FOR QUASI-STATIC PLANE STRAIN DEFORMATIONS OF GRANULAR MATERIAL

IAN F. COLLINS*

Abstract. An analysis of the quasi-static, plane strain, deformation of a rigid/plastic material whose yield stress depends on the current material density is presented. Such constitutive equations are widely used to model the initial yielding, final failure and flow of granular materials in what is generally known as "critical state soil mechanics". It will be shown that the mathematical structure of these problems is rather more complex than seems to have been hitherto realized. It will be shown that there exist families of weak discontinuities other than "the stress and velocity characteristics" and that they provide a key to the resolution of the long standing debate regarding the non-coincidence of stress and velocity characteristics for frictional materials. The rôle of isotropy, anisotropy, normal flow rules and non-normal flow rules will be discussed.

1. Introduction. The theory of plane strain deformations of plastically deforming materials has been extensively studied. One reason for the concentration on two dimensional problems is that the equations are almost invariably hyperbolic, unlike the governing equations for three-dimensional deformations which are usually of mixed type. The theory of plane strain, quasi-static, deformations of rigid/plastic models of metals has reached a particularly highly developed state [6,7], but the same cannot be said for rigid/plastic models for soils and granular materials. Whilst much initial progress was made in the study of the *statics* of such problems - as exemplified by the large number of "stress solutions" discussed in the book by Sokolovskii [29], it has proved difficult to develop a universally acceptable theory of the *kinematics* of the deformations associated with the failure and flow of soils and granular materials. The use of a "normal flow rule", so widely and successfully employed in conventional metal plasticity, in conjunction with the fundamental linear Coulomb failure criterion leads to predictions of volume changes which are far larger than any actually observed in practice. This fact has lead many authors to adopt a non-normal flow rule, which can lead to problems of lack of uniqueness and non-coincidence of stress and velocity characteristics as explained in Section 3. Other workers have constructed models which preserve the normality rule but use nonlinear yield criteria and incorporate strain hardening and softening.

The object of the present paper is to compare the mathematical structure of these two classes of theories, and to demonstrate that many of the difficulties associated with the non-coincidence of stress and velocity characteristics disappear when the dependence of the yield condition on the material density is properly allowed for.

*Department of Engineering Science, University of Auckland, Auckland, New Zealand

2. Elements of a Theory of Plasticity. The constitutive equations for a plastically deforming material consist of a yield condition, flow rule and hardening rule. A yield condition

$$f(\sigma_{ij}) = 0 \tag{2.1}$$

can be thought of as defining a convex surface in six dimensional stress space. If the point in this space representing the stress σ_{ij} in a material element lies inside this surface, this material element will deform elastically or remain rigid if, as will be done throughout this paper, elastic strains are neglected. Only when the stress point reaches the yield surface can irreversible plastic strains occur. If the material is isotropic the yield function f will only depend on the three stress invariants. The manner in which a material element deforms is given by the flow rule, which is a relation between the stress σ_{ij} and strain rate e_{ij} (rate of deformation) tensors such as

$$e_{ij} = \lambda \partial f / \partial \sigma_{ij}. \tag{2.2}$$

The particular form (2.2) is called a "normal" or "associated flow rule". It has the geometric interpretation that the six-dimensional vector representing the strain-rate components is directed along the outward normal to the yield surface at the current stress-point (Fig. 1). The quantity λ is simply a proportionality factor, it is not a material constant. The flow rule (2.2) requires the components of the strain-rate and yield function gradient tensors to be proportional. This reflects the fact that the deformation is rate independent. The normal flow rule is widely used in metal plasticity, where its use can be justified by averaging arguments and the slip properties of individual crystals. This rule also ensures that the stress/strain-rate relation is convex and hence provides sufficient conditions for uniqueness [6]. The simplest class of "non-normal flow rules" are obtained by replacing the yield function f in (2.2) by another distinct "plastic potential" function, $g(\sigma_{ij})$ say. More complicated flow rules, which involve other tensors such as stress-rates are of course possible - we shall refer to these rules as being "non-associated".

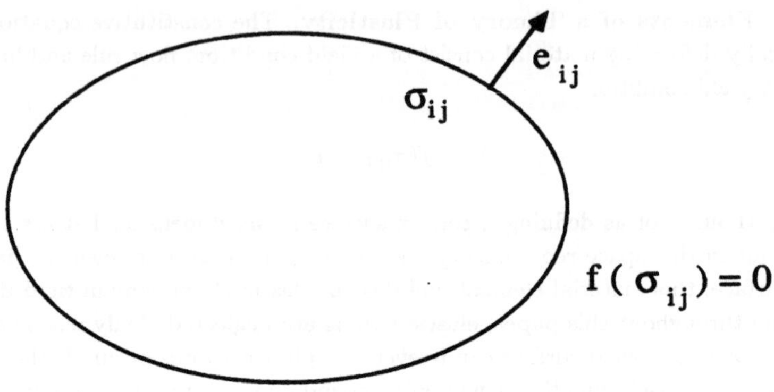

FIG 1 - YIELD SURFACE AND NORMAL FLOW RULE

In a "perfectly" plastic material the yield surface remains unaltered as the deformation proceeds. In practice the yield surface is frequently found to expand (isotropic hardening) or contract (isotropic softening) and translate (kinematic hardening), and a variety of models for soil behaviour have been proposed which include these effects, (see Section 4).

In the classical plane strain theory of the perfectly plastic behaviour of metals, yielding occurs when the maximum shear stress reaches a critical value k [14]:

$$(\sigma_{11} - \sigma_{22})^2/4 + \sigma_{12}^2 = k^2. \tag{2.3}$$

For a given material element the shear stress is a maximum on line segments which bisect the principal directions. The trajectories of these maximum shear stress directions are called sliplines, which hence form a two parameter orthogonal network of curves in the plane of deformation.

The (quasi-static) equilibrium equations are

$$\begin{aligned} \sigma_{11,1} + \sigma_{12,2} &= 0 \\ \sigma_{12,1} + \sigma_{22,2} &= 0 \end{aligned} \tag{2.4}$$

which when combined with (2.3) form a statically determinate set of 3 equations for 3 unknowns. It is easily shown that these equations are hyperbolic, with the sliplines as characteristic [6,14].

The corresponding velocity equations obtained from the normal flow rule (2.2) are

$$(2.5)\qquad \begin{cases} v_{1,1} = e_{11} = \dfrac{1}{2}\,\lambda(\sigma_{11} - \sigma_{22}) \\[2ex] v_{2,2} = e_{22} = -\dfrac{1}{2}\,\lambda(\sigma_{11} - \sigma_{22}) \\[2ex] \dfrac{1}{2}\,(v_{1,2} + v_{2,1}) = e_{12} = \lambda\sigma_{12} \end{cases}$$

If the reference axes are taken locally parallel to the slipline directions it is easily shown that $\sigma_{11} = \sigma_{22}$, so that (2.5) reduce to:

$$(2.6)\qquad v_{1,1} = v_{2,2} = 0.$$

Since these are two ordinary differential relations the sliplines must also be characteristics of the velocity equations, which are also seen to be the directions of zero extension-rate.

3. Perfectly Plastic Models for Granular Materials. Early attempts at generalising the above theory to soils and granular media soon ran into serious difficulties. The simplest yield condition which models the failure behaviour of granular materials under plane strain conditions is the linear Coulomb condition:

$$(3.1)\qquad |\tau| = c + \sigma \tan\psi$$

where (σ, τ) are the normal (compressive) and shear traction on a material element, and c the cohesion and ψ the angle of internal friction are material parameters. In terms of stress invariants the condition can be rewritten

$$(3.2)\qquad q = p \sin\psi + c \cos\psi$$

where

$$(3.3)\qquad p = \frac{1}{2}\,(\sigma_{11} + \sigma_{22}),\; q = \left[\frac{1}{4}\,(\sigma_{11} - \sigma_{22})^2 + \sigma_{12}^2\right]^{1/2}$$

are the mean pressure and maximum shear stress respectively (N.B. compressive stresses and strain-rates will be taken as positive). This condition reduces to the Mises of Tresca failure condition (2.3) when $\psi = 0$.

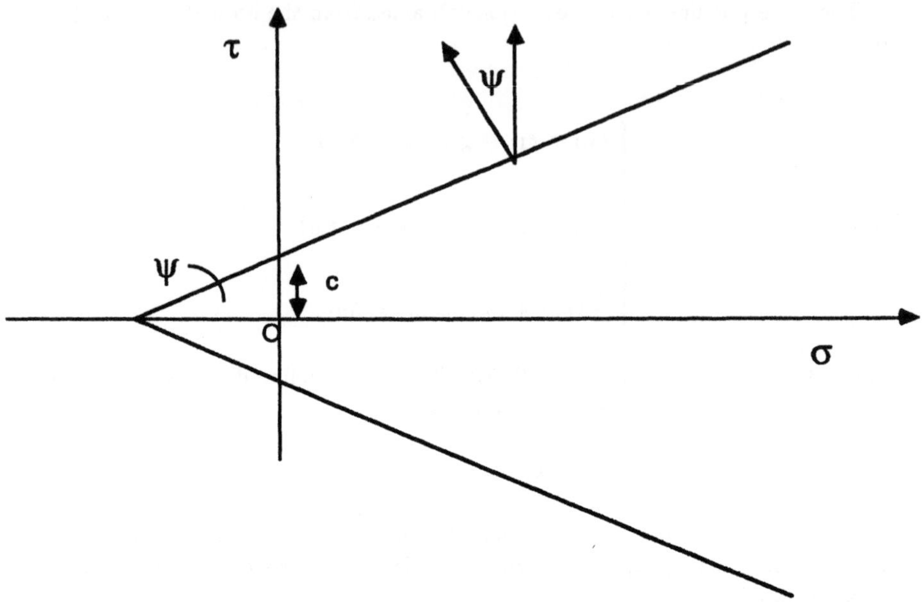

FIG 2 - COULOMB FAILURE CONDITION

Elementary stress analysis shows that condition (3.1) is attained on line segments which make an angle of $\pm(\pi/4 - \psi/2)$ with the more compressive principal stress, and that the corresponding curvilinear trajectories are the stress characteristics, which are such that when used as co-ordinate lines, the equilibrium equations can be written as a pair of ordinary differential equations. Sokolovskii gives a comprehensive account of this theory in his book [29] and gives a number of "solutions" to particular boundary value problems of importance in geotechnical engineering. However, since no attempt is made to associate deformation fields with these stress fields, they must be viewed as only partial solutions.

The theory of the kinematics of the flow field obtained by using a normal flow rule with the Coulomb yield condition was developed by Shield [28], who found that the characteristics of the velocity field coincided with the Sokolovskii stress characteristics. However, as can be seen from Fig. 2 this normal flow rule predicts strain-rates with a dilatational component. It is found in practice that this dilational component far exceeds those actually observed in practice.

Hill [14] suggested an alternative model in which the material is assumed to be incompressible, so that the strain-rate vector in the yield diagram, Fig. 2, is now parallel to the τ-axis. Since the material is still presumed isotropic the principal axes of stress and strain-rate are made to coincide. Consequent on these two assumptions it is now found that the velocity and stress characteristics do not coincide, the former being at angles of $\pm\pi/4$ to the major principal axis. This model hence has the undesirable feature that whilst the failure condition is met on line segments parallel to the Sokolovskii stress characteristics, the material actually fails by shearing along the distinct velocity characteristics.

A number of theories have been proposed to overcome these difficulties. These fall into two classes. In one, the material is still assumed to be perfectly plastic, but more complex flow rules are employed to overcome the above problems, whilst the second class introduces work-hardening and/or softening into the constitutive model. The latter class of models is described in the next Section.

The basic concept underlying these extended perfectly plastic models is that the material deforms by sliding or shearing along one or both of the Sokolovskii stress characteristics. Geniev [13] and Mandl and Fernandez Luque [19] assumed that the material failed by shearing along one of the stress characteristics, whilst Mandel [18], Spencer [30], Mehrabadi and Cowin [20] and Mróz and Szymánski [23] and others have developed a theory in which the deformation is the sum of two shearing motions, one along each characteristic plus a rigid body spin, which is chosen so as to make the stress and velocity characteristics coincide. The flow rule in these models is a relation between three tensors; stress, strain-rate and stress-rate, so that the assumption that the material is isotropic does not imply that the principal axes of stress and strain-rate coincide. This is why it is possible to reconcile the abandonment of a normal flow rule with the assumption of material isotropy, but still have the velocity and stress characteristics coinciding. In Spencer's original theory [30] the material was assumed to be incompressible. The generalisation to dilatant materials was made in [20].

De Josselin de Jong [9,10] has developed a similar theory but in which the spin is not restricted and the two sets of characteristics do not coincide, this leads to a constitutive theory which is incomplete, though uniqueness can be obtained by appealing to minimum energy hypotheses. General reviews of these perfectly plastic theories can be found in the following articles [5,12,15,16,21,22,31].

One of the prime objectives of the present paper is to show that when the Coulomb line is viewed as the so called critical state line, which represents the ultimate failure state in a density hardening model, the distinction between these stress and velocity characteristics occurs in a natural manner, so that the need to introduce non-associated flow rules is removed.

4. Hardening Models and the Critical State Concept. There have been many attempts at introducing strain hardening to more realistically describe the physical behaviour of soils [11,25]. Unlike most comparable models for non-porous metals, the hardening parameters for granular materials depend on volumetric strains. The basic constitutive model which includes dilational hardening/softening is now known as critical state soil mechanics [1,2,3,4,16,27]. The material is assumed to be isotropic, but with a yield function which depends on the density ρ:

$$F(p,q,\rho)=0 \tag{4.1}$$

(Note: for simplicity of presentation in this paper the presence of pore water and pore pressures will be ignored so the theory is strictly only applicable to dry sands.

However, the main arguments carry over unchanged when pore water is included as explained in [8], the main difference being that the density must be replaced as a state variable by the specific volume).

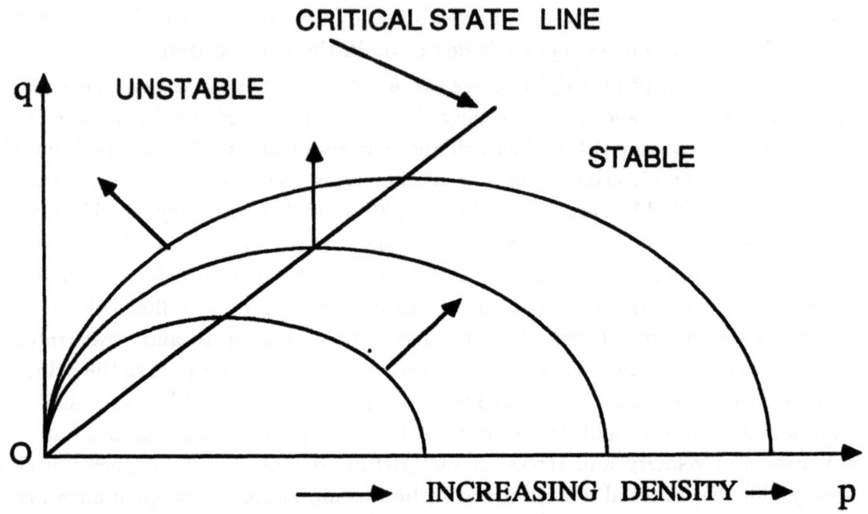

FIG 3 - CRITICAL STATE MODEL

This function can be represented by a "state boundary surface" in p, q, ρ space or by a nest of curves in a (q,p) diagram as shown in Fig. 3. Several slightly different forms of this function are used in the literature [1,16,24] but they all have the property of being convex with the density increasing outwards, and such that their tangent is horizontal on the straight line - the so called critical state line:

(4.2) $$q = p \sin \psi_c$$

where ψ_c is a material constant. This equation is the same as that for a cohesionless Coulomb material with ψ_c as the angle of internal friction.

If the normal flow rule is assumed, the strain-rate corresponding to a stress state to the right of the critical state line will have a compressive component, so that the material density will increase and, provided the material element remains plastic it hardens so that the point representing the current stress state must move to an outer yield surface. On the other hand a material element whose stress-state is to the left of critical, will have a dilational strain-rate component, so that its density will decrease and the material will soften. This is an unstable process and such deformations tend to be localized into shear bands - such behaviour is typical of dense sands and overconsolidated clays. Conversely the deformations occuring on the right hand side of critical will be stable and continuous, the overall deformation being confined until the stress-path reaches the critical state line. This is typical of the behaviour of loose sands and normally consolidated clays. (N.B.

This conventional view may have to be modified in the light of the recent results of Pitman and Schaeffer [24], who have shown that instabilities can occur even on the right hand side of the critical state line.)

Since the strain-rate vector is parallel to the q-axis for a stress-point actually on the critical state line, the corresponding deformations are isochoric. Velocity discontinuities which occur, for example at the failure of a slope, cannot occur in the stable regime, since any tendency to produce an intense band of shear is accompanied by a corresponding compressive strain-rate. Only when the stress state reaches the critical state line can gross deformations occur. The critical state line represents the ultimate failure or full flow condition for materials whose initial states are on the stable side of this line. In this sense the critical state line is playing the rôle of the Coulomb failure line in the perfectly plastic models described in the last section.

The critical state model should perhaps be viewed as the second in the hierarchy of constitutive models, the simple Coulomb material being the first. It would seem to work well for clays but is less appropriate for describing the behaviour of sands [17].
Various generalisations have been proposed which include induced anisotropy, shear in addition to dilatational hardening, non-associated flow rules and yield surfaces containing vertices. These extensions are needed to explain observed phenomena, particularly the formation of shear bands in shear box and biaxial tests.

5. Characteristics and Weak Discontinuities.

5.1 Weak Discontinuities in Stress. In order to understand the characteristic structure of the governing equations of critical state models, we here investigate the possibility of the derivatives of the dependent variables - stress σ, velocity v and density ρ being discontinuous across a curve $\mathcal{C}$ say. The stress components satisfy the equilibrium equations

$$(5.1) \qquad \sigma_{tt,t} + \sigma_{tn,n} = \rho X_t, \qquad \sigma_{tn,t} + \sigma_{nn,n} = \rho X_n,$$

where t and n denote components locally tangential and normal to $\mathcal{C}$ and (X_t, X_n) are the components of the local body force density, together with the state boundary surface condition

$$(5.2) \qquad f(\sigma_{tt}, \sigma_{nn}, \sigma_{nt}, \sigma_{tn}, \rho) = 0$$

which is of course symmetric in σ_{nt} and σ_{tn}, or if the material is isotropic, as in the classical critical state models:

$$(5.3) \qquad F(p, q, \rho) = 0$$

Since we are concerned here with weak discontinuities, the stress, velocity and density must all be continuous across $\mathcal{C}$ as must therefore also be their tangential derivatives. It follows from (5.1) that only the values of the normal derivative of σ_{tt} can possibly be discontinuous.

Differentiating the state boundary function condition (5.2) in the n-direction gives

$$(5.4)\qquad f_{,tt}\,\sigma_{tt,n} + 2f_{,tn}\,\sigma_{tn,n} + f_{,nn}\,\sigma_{nn,n} + f_{,\rho}\,\rho_{,n} = 0$$

where $f_{,tt} \equiv \partial f/\partial\sigma_{tt}$, etc.

This equation holds on both sides of $\mathfrak{C}$ provided the material on each side is stressed up to yield. Hence any jump in $\sigma_{tt,n}$ must satisfy the condition

$$(5.5)\qquad f_{,tt}\,[\sigma_{tt,n}] + f_{,\rho}\,[\rho_{,n}] = 0$$

where $[\cdot]$ denotes the jump in the argument. In practice the state boundary function f is such that $f_{,\rho}$ is never zero. There are hence two possibilities:

$$(a)(5.6)\qquad f_{,tt} = 0,$$

in which case $[\rho_{,n}]$ must be zero, but a jump in the normal derivative of σ_{tt} is possible. The first term in (5.4) is now zero and the normal-derivatives of the stress components in the second and third terms can be expressed in terms of tangential-derivatives using (5.1). It follows therefore that (5.4) can be rewritten as an *ordinary* differential relation along $\mathfrak{C}$, which is hence a characteristic of the stress equations. The form of this equation for an isotropic material is developed in [8]. Curves of this type we shall denote by $\mathfrak{C}_1$.

(b) If $f_{,tt} \neq 0$, it follows from (5.5) that any jump in $\sigma_{tt,n}$ must be accompanied by a jump in the density gradient. Furthermore since the first term in (5.4) does not vanish, this second class of curve, denoted by $\mathfrak{C}_2$, is not a stress characteristic, it should perhaps best be viewed as a density wave.

5.2 Weak Discontinuities in Velocity. The velocity components are given by the flow rule

$$(5.7)\qquad -v_{t,t} = \lambda g_{,tt}$$

$$(5.8)\qquad -v_{n,n} = \lambda g_{,nn}$$

$$(5.9)\qquad -1/2(v_{t,n} + v_{n,t}) = \lambda g_{,tn} = \lambda g_{,nt}$$

where $g(\sigma_{ij}, \rho)$ is the plastic potential, which specialises to $G(p, q, \rho)$ for an isotropic material. In addition the velocity components and density are related through the continuity equation:

$$(5.10)\qquad \rho_{,t} + \rho(v_{t,t} + v_{n,n}) + v_t\rho_{,t} + v_n\rho_{,n}) = 0.$$

If $\mathfrak{C}$ is a weak discontinuity in the velocity field, we observe from (5.8) and (5.9) that the normal derivatives of velocity can only be discontinuous if $[\lambda] \neq 0$. But from (5.7) this can only happen when both

$$(5.11)\qquad v_{t,t} = 0 \text{ and } g_{,tt} = 0.$$

The first part of (5.11) is an *ordinary* differential relation showing that $\mathcal{C}$ is a characteristic of the velocity equations, and that it is a zero extension-rate direction. The second part of (5.11) gives the direction of $\mathcal{C}$ relative to the principal axes of stress - this will depend on the precise form of the plastic potential function. It follows from (5.6) and (5.11) that the stress and velocity characteristics will coincide for a material with a normal flow rule.

The continuity equation (5.10) puts further restrictions on the allowable jumps in the velocity and density gradients. Since the tangential derivatives must be continuous across $\mathcal{C}$ it follows from (5.10) that

$$[\rho_{,t}] + \rho[v_{n,n}] + v_n[\rho_{,n}] = 0. \tag{5.12}$$

However, since ρ remains continuous across the moving surface, these jumps also satisfy the kinematic relation [8]:

$$[\rho_{,t}] + V[\rho_{,n}] = 0 \tag{5.13}$$

where V is the speed of the discontinuity. Eliminating the jump in the time derivative of ρ between (5.12) and (5.13) gives

$$\rho[v_{n,n}] = (V - v_n)[\rho_{,n}] \tag{5.14}$$

5.3 Normal and Non-Normal Flow Rules. Consider again the type $\mathcal{C}_1$ weak discontinuities discussed above. These are stress characteristics, in the direction given by (5.6), which admit discontinuities in stress gradients but not density gradients. It follows therefore from (5.14) that the velocity gradients and hence the strain-rate must be continuous across a $\mathcal{C}_1$ curve. For a material with a non-normal flow rule, this is of no consequence, since the velocity and stress characteristics do not coincide anyway. For a standard material with a normal flow rule however, it shows that although the stress and velocity characteristics coincide, only the gradient of stress can be discontinuous across such a curve, the gradients of both velocity and density are continuous. (This point was not brought out correctly in [8]). The *second* gradients of velocity can be discontinuous however, and these jumps can be related to the stress gradients by differentiating the flow rule (5.7 - 5.9) in the normal direction. (N.B. These arguments must be modified if the characteristic is a loading or unloading boundary with rigid material on one side [8].)

We turn now to the second type of weak discontinuity $\mathcal{C}_2$, which admits concurrent jumps in stress and density gradient which are related by (5.5). This curve is not a stress characteristic and hence in a standard material cannot be a velocity characteristic either, so that the velocity gradient must be continuous across $\mathcal{C}_2$. The jump conditions (5.5) and (5.14) are hence only compatible if $v_n = V$, so that $\mathcal{C}_2$ must be a ***contact discontinuity***, embedded in the material.

The situation is rather different for a material deforming according to a non-normal flow rule. Since the stress and velocity characteristics no longer coincide there are now two possibilities.

(i) If $\mathfrak{C}_2$ is also not a velocity characteristic, then the above argument applies and the curve must be an embedded contact discontinuity.

(ii) When $\mathfrak{C}_2$ is also a velocity characteristic, there can now be jumps in the gradient of velocity as well as density and stress, the three jumps being related by (5.5) and (5.14) and it is *not* now necessary for such a jump to be embedded in the material. This type of *moving* weak discontinuity we shall denote by $\mathfrak{C}_2'$.

Table 1: SUMMARY OF CLASSES OF WEAK DISCONTINUITIES IN DENSITY DEPENDENT RIGID/PLASTIC MATERIALS

Curve	Normal Flow Rule ($g = f$)	Non-Normal Flow Rule ($g \neq f$)
$\mathfrak{C}_1$	**Direction:** $f_{,tt} = 0$ $v_{t,t} = 0$ zero extension rate direction Coincident stress and velocity characteristics **Discontinuities:** $[\sigma_{tt,n}]$ **Continuous:** $\rho_{,n}$ and $\mathbf{v}_{,n}$ **In critical state limit** $\mathfrak{C}_1 \to$ orthogonal curves and $[\sigma_{tt,n}] \to 0$.	**Direction:** $f_{,tt} = 0$ Stress, but not velocity characteristics **Discontinuities:** $[\sigma_{tt,n}]$ **Continuous:** $\rho_{,n}$ and $\mathbf{v}_{,n}$ **In critical state limit** $\mathfrak{C}_1 \to$ non-orthogonal curves and $[\sigma_{tt,n}] \to 0$.
$\mathfrak{C}_2$	**Direction:** depends on magnitude of jumps **Discontinuities:** $[\sigma_{tt,n}], [\rho_{,n}]$ **Continuous:** $\mathbf{v}_{,n}$ Embedded contact discontinuities$\to$ $\mathfrak{C}_3$ "Sokolovskii characteristics" in **critical state limit**	
$\mathfrak{C}_2'$		**Direction:** $g_{,tt} = 0$ $v_{t,t} = 0$ zero extension-rate direction Velocity, but not stress characteristics **Discontinuities:** $[\sigma_{,tt,n}][\rho_{,n}]$ and $[\mathbf{v}_{,n}]$ Discontinuity travels with speed V given by (5.14). in **critical state limit** $\mathfrak{C}_2' \to \mathfrak{C}_3$ "Sokolovskii characteristics"

The results established in this section are summarized in Table 1. It is of interest to note that it is only possible to have an "active" velocity characteristic - i.e. one with a discontinuous velocity gradient in a material with a non-normal flow rule. For a standard material any jump in velocity gradient, would have be accompanied by one in density gradient from (5.14), which is impossible from (5.5) and (5.6) when the stress and velocity characteristics coincide.

5.4 The Critical State Limit. For a general anisotropic material with a non-normal flow rule the "critical state" is a family of stress states obtained by eliminating the density between the state boundary surface condition (5.2) and the condition that the material is deforming without further change in volume:

(5.15) $$\partial g/\partial \sigma_{ii} = 0 \qquad \text{(summation convention)}.$$

For a standard isotropic material, this family of stress-states consists of the critical state line (4.2). For the general material we shall denote these critical stress states by:

(5.16) $$\tilde{f}(\sigma_{ij}) \equiv f(\sigma_{ij}, h(\sigma_{ij})) = 0$$

where (5.15) has been rewritten in the form

(5.15′) $$\rho = h(\sigma_{ij}).$$

If all the material elements in a certain region of a deforming body have reached the critical state the stresses will satisfy the equilibrium equations (5.1) and the critical state condition (5.16). Arguing as above we see that the stress-characteristics of this system are in the directions

(5.17) $$\tilde{f}_{,tt} \equiv f_{,tt} + f_{,\rho}\, h_{,tt} = 0, \text{ using } (5.15)' \text{ and } (5.16).$$

These curves, which are in general distinct from the $\mathfrak{C}_1$ curves discussed above, will be denoted by $\mathfrak{C}_3$. It is important to appreciate that in the critical state limit there are *two* distinct families of stress characteristics: (i) $\mathfrak{C}_1$-characteristics which arise because the stress-state lies on the state boundary surface (5.2) and (ii) $\mathfrak{C}_3$-characteristics which exist because the stress-state also satisfies the critical state condition (5.16). For an isotropic material, with (4.2) as the critical state line, these $\mathfrak{C}_3$ curves are the Sokolovskii characteristics.

Note that it follows from (5.15)′ that any jump in density and stress gradients in a critical state region must satisfy:

(5.18) $$[\rho_{,n}] = h_{,tt}\, [\sigma_{tt,n}],$$

so that since ρ is continuous across a $\mathfrak{C}_1$ curve (Table 1), it follows that in the critical state limit the jump in stress gradient across such a curve tends to zero. In this sense therefore the $\mathcal{C}_1$ stress characteristics have vanishing strength in the critical state limit.

Eliminating the jump in the density gradient between (5.5) and (5.18) shows that in the critical state limit, the jump in stress gradient across a curve of class $\mathcal{C}_2$ satisfies

$$(5.19) \qquad (f_{,tt} + f_{,\rho}\, g_{,tt})[\sigma_{tt,n}] = 0$$

so that if this jump is non-zero, it must occur across a curve whose direction satisfies (5.17). In other words in the critical state limit the $\mathcal{C}_2$ discontinuities become the $\mathcal{C}_3$ stress characteristics. This generalizes the results established in [8] for an isotropic material with a normal flow rule, to a general anisotropic material with either a normal or non-normal flow rule.

Discussion and Conclusions. It has been shown that there are in general two types of possible weak discontinuity which can occur in the quasi-static, plane strain deformation of a rigid/plastic material whose yield function depends on the current density. One class ($\mathcal{C}_1$) are the characteristics of the stress or velocity equations in the sense that they admit ordinary differential relations in the direction of the discontinuities. The existence of these curves is well known and can be demonstrated using the standard techniques of characteristics theory [16,24,26]. However, the existence of the second class ($\mathcal{C}_2$) of weak discontinuities does not seem to be well known, at least in the present context. They involve simultaneous jumps in the gradients of two variables (stress and density) and hence are not characteristics in the interior derivative sense.

A state of stress lying in the critical state region satisfies *two* algebraic constraints. It lies on a particular yield curve (5.2) or (5.3) but the stress tensor also satisfies (5.16) since the stress state lies in the critical state region. There are hence *two* families of characteristics corresponding to these two constraints. The first family, whose directions are given by (5.6) are the critical state limits of the $\mathcal{C}_1$ characteristics. The second family of (Sokolovskii) stress characteristics are the $\mathcal{C}_3$ curves given by (5.17) and these are the critical state limits of the $\mathcal{C}_2$ density waves. The presence of these two families of stress characteristics is not recognised in the *direct* analyses of the perfectly plastic models reviewed here in Section 3 - the $\mathcal{C}_3$ curves are the only characteristics to appear. This fact has lead, in the author's view, to the unnecessary complication of the theory by the introduction of non-associated flow rules. The present analysis has shown that for a normal flow rule material in the critical state the velocity characteristics coincide with the $\mathcal{C}_1$ stress characteristics, so that the model possesses "the desirable property", discussed in Section 3, of the material element failing by shearing in the directions on which the stresses satisfy the failure condition. The theories [18,20,30] which introduce non-associated flow rules, involving stress-rates, force the velocity characteristics to coincide with the $\mathcal{C}_3$ stress characteristics, which as has been seen are the limits of density waves and not of any form of velocity characteristics. The difference in the two view points arises, of course, because in the hardening model the Coulomb line is viewed as an asymptotic failure state, whilst in the perfectly plastic models it is the yield curve.

REFERENCES

[1] J.H. Atkinson, *Foundation and Slopes*, McGraw-Hill, London (1981).

[2] J.H. Atkinson, and P.L. Bransby, *The Mechanics of Soils*, McGraw-Hill, London (1978).

[3] M. Bolton, *A Guide to Soil Mechanics*, Macmillan, London (1979).

[4] A.M. Britto and M.J. Gunn, *Critical State Soil Mechanics via Finite Elements*, Ellis Harwood, Chichister (1987).

[5] R. Butterfield and R.M. Harkness, *The Kinematics of Mohr-Coulomb Materials*, Proc. Roscoe Memorial Symposium "Stress-Strain Behaviour of Soils" ed. by R.H.G. Parry, G.T. Foulis Henley (1972), pp. 220–233.

[6] J. Chakarabarty, *Theory of Plasticity*, McGraw-Hill, London (1987).

[7] I.F. Collins, *Boundary Value problems in plane strain plasticity*, Mechanics of Solids, the Rodney Hill 60th Anniversary Volume, eds H.G. Hopkins and M.J. Sewell, Pergamon Press, Oxford (1981), pp. 135–184.

[8] ————, *Plane strain characteristics theory for soils and granular materials with density dependent yield criteria*, J. Mech. Phys. Solids (in press)..

[9] G. De Josselin De Jong, *The double-sliding, free-rotating model for granular assemblies*, Géotechnique, 21 (1971), pp. 155–163.

[10] ————, *Mathematical elaboration of the double-sliding, free rotating model., Archs. Mech., 29 (1977), pp. 561–591.*

[11] D.C. Drucker, R.E. Gibson and D.J. Henkel, *Soil mechanics and work-hardening theories of plasticity*, Trans A.S.C.E. 122 (1957), pp. 338–346.

[12] E.H. Davis and J.R. Booker, *Some adaptions of classical plasticity theory for soil stability problems*, Symposium on Plasticity and Soil Mechanics, Cambridge, (1973), pp. 24–41.

[13] G.A. Geniev, *Problems of dynamics of granular media*, Akad. Stroit. i. Arch. Moscow (1959).

[14] R. Hill, *The Mathematical Theory of Plasticity*, Clarendon Press, Oxford (1950).

[15] G.T. Houlsby and C.P. Wroth, *Direct solution of plasticity problems in soils by the method of characteristics*, Proc. 4th Int. Conf. Num. Methods in Geomechanics, Edmonton, Canada, 2 (1982), pp. 1059–1071.

[16] R. Jackson, *Some mathematical and physical aspects of continuum models for the motion of granular materials*, Theory of Dispersed Multiphase flow, ed. R.E. Meyer, Academic Press (1983), pp. 291–337.

[17] P.V. Lade, *Effects of voids and volume changes on the behaviour of frictional materials*, Int. J. Num. and Anal. Methods Geomechanics, 12 (1988), pp. 351–370.

[18] J. Mandel, *Sur les équations à écoulement des sols idéaux en déformations plane et le concept de double glissement*, J. Mech. Phys. Solids, 14 (1966), pp. 303–306.

[19] G. Mandl and R. Fernandez Luque, *Fully developed plastic shear flow of granular materials*, Géotechnique, 20 (1970), pp. 277–307.

[20] M.M. Mehrabadi and S.C. Cowin, *Initial planar deformation of dilatant granular materials*, J. Mech. Phys. Solids, 26 (1978), pp. 268–284.

[21] ————, *On the double sliding free-rotating model for the deformation of granular materials*, J. Mech. Phys. Solids, 29 (1981), pp. 269–282.

[22] Z. Mroz, Proc. 15th IUTAM Congress, Toronto, ed. F.P.J. Rimrott and B. Tabarrok, North Holland, Amsterdam (1980), pp. 119–132, *Deformation and flow of granular materials.*

[23] Z. Mroz and CZ. Szymanski, *Non-associated flow rules in description of plastic flow of granular materials*, in "Limit analysis and rheological approach to soil mechanics", CISM Course 217, Udine (1979), pp. 50–94.

[24] D.B. Pitman and D.G. Schaeffer, *Stability of time dependent compressible granular flow in two dimensions*, Comm. Pure Appl. Maths., 40 (1987), pp. 421–447.

[25] K.H. Roscoe, A.N. Schofield and C.P. Wroth, *On the yielding of soils*, Géotechnique, 8 (1958), pp. 22–53.

[26] D.G. Schaeffer, *Instability in the evolution equations describing incompressible granular flow*, J. Diff. Equations 66 (1987), pp. 19–50.

[27] A.N. Schofield and C.P. Wroth, *Critical State Soil Mechanics, McGraw-Hill, London (1968).*

[28] R.T. Shield, *Mixed boundary value problems in soil mechanics*, Q. Appl. Math., 11 (1953), pp. 61–75.

[29] V.V. Sokolovskii, *Statics of Granular Media*, Pergamon Press, Oxford (1965).

[30] A.J.M. Spencer, *A theory of the kinematics of ideal soils under plane strain conditions*, J. Mech. Phys. Solids, 12 (1964), pp. 337–351.

[31] ————————, *Deformation of ideal granular materials*, Mechanics of Solids, The Rodney Hill 60th Anniversary Volume, eds. H.G. Hopkins and M.J. Sewell, Pergamon Press, Oxford (1981), pp. 607–652.

RELATION OF MICROSTRUCTURE TO CONSTITUTIVE EQUATIONS

DONALD A. DREW,* GARY S. ARNOLD† AND RICHARD T. LAHEY, JR.‡

Abstract. A set of equations for the averaged motion of a suspension of spheres in an inviscid fluid is derived by applying an ensemble average to the exact equations of motion. Constitutive equations to close the system for dilute, incompressible flow are derived by averaging the solution for the flow around a single sphere. The ensemble average is performed by allowing the center of the sphere to take random locations around an arbitrary spatial point. The corresponding volume average yields different results.

1. Exact Equations. The exact equations of motion, valid inside each material are

Mass

$$\frac{\partial \rho}{\partial t} + \nabla \cdot \rho \mathbf{v} = 0 \tag{1}$$

Momentum

$$\frac{\partial \rho \mathbf{v}}{\partial t} + \nabla \cdot \rho \mathbf{v}\mathbf{v} = \nabla \cdot \mathbf{T} + \rho \mathbf{g} \tag{2}$$

where ρ is the density, $\mathbf{v}$ is the velocity, $\mathbf{T}$ is the stress tensor

$$\mathbf{T} = -p\mathbf{I} + \boldsymbol{\tau} \tag{3}$$

where p is the pressure and $\boldsymbol{\tau}$ is the shear stress.

For non-polar materials, conservation of angular momentum implies

$$\mathbf{T} = \mathbf{T}^t. \tag{4}$$

2. Ensemble Averaging. Averaging involves simply adding the observed values and dividing by the number of observations. The appropriate general average is the ensemble average.

If f is some field (i.e., a function of position $\mathbf{x}$ and time t) for some particular realization, μ, of the process, then the average of f is

$$\overline{f}(\mathbf{x}, t) = \int_M f(\mathbf{x}, t; \mu) dm(\mu) \tag{5}$$

---

*Department of Mathematical Sciences, Rensselaer Polytechnic Institute, Troy, NY 12180-3590, on Sabbatical Leave with Mathematical Sciences Institute, Cornell University, Ithaca, NY 14853, from 1 September 1988 to 30 June 1989.

†General Electric Company, 1 River Road, Schenectady, NY 12301

‡Department of Nuclear Engineering and Engineering Physics, Rensselaer Polytechnic Institute, Troy, NY 12180-3590

where $dm(\mu)$ is the measure (probability) of observing process μ and M is the set of all processes.

In order to average to the exact equations, we need expressions for $\overline{\partial f/\partial t}$ and $\overline{\nabla f}$. If f is "well behaved", then it is clear from the definition of the ensemble average that

$$\overline{\frac{\partial f}{\partial t}} = \frac{\partial \overline{f}}{\partial t} \tag{6}$$

and

$$\overline{\nabla f} = \nabla \overline{f} \tag{7}$$

Functions are generally discontinuous at the interface in most multiphase flow. They are well behaved within each phase, however. Thus, consider $\overline{X_k \nabla f}$, where X_k is the phase indicator function for phase k:

$$X_k = \begin{cases} 1, & \text{if } \mathbf{x} \in k; \\ 0, & \text{otherwise.} \end{cases} \tag{8}$$

Then

$$\begin{aligned} \overline{X_k \nabla f} &= \overline{\nabla X_k f} - \overline{f \nabla X_k} \\ &= \nabla \overline{X_k f} - \overline{f \nabla X_k}. \end{aligned} \tag{9}$$

and

$$\overline{\frac{X_k \partial f}{\partial t}} = \overline{\frac{\partial X_k f}{\partial t}} - \overline{f\frac{\partial X_k}{\partial t}} = \frac{\partial \overline{X_k f}}{\partial t} - \overline{\frac{f \partial X_k}{\partial t}}. \tag{10}$$

The second term on the right hand side in both of these equations is related to the surface average of f, evaluated on the k-phase side, over the interface. We note that

$$\frac{\partial X_k}{\partial t} + \mathbf{v}_i \cdot \nabla X_k = 0$$

where $\mathbf{v}_i$ is the velocity of the interface.

$$f - \overline{f\frac{\partial X_k}{\partial t}}$$

$$= \frac{\partial \overline{X_k f}}{\partial t} - \overline{\frac{f \partial X_k}{\partial t}}. \tag{10}$$

3. Averaged Equations. The averaged equations are

Mass

$$\frac{\partial \overline{X_k \rho}}{\partial t} + \nabla \cdot \overline{X_k \rho \mathbf{v}} = \overline{\rho(\mathbf{v} - \mathbf{v}_i) \cdot \nabla X_k} \tag{11}$$

Momentum

$$\frac{\partial \overline{X_k \rho \mathbf{v}}}{\partial t} + \nabla \cdot \overline{X_k \rho \mathbf{v}\mathbf{v}} = \nabla \cdot \overline{X_k \mathbf{T}} + \overline{X_k \rho \mathbf{g}} + \overline{(\rho \mathbf{v}(\mathbf{v} - \mathbf{v}_i) - \mathbf{T}) \cdot \nabla X_k}. \tag{12}$$

The volume fraction is defined as

$$\alpha_k = \overline{X_k} \tag{13}$$

All the remaining variables are defined in terms of weighted averages. The main, or "phasic" variables are either phasic weighted variables (weighted with the phase function X_k) or mass-weighted (or Favré) averaged (weighted by $X_k \rho$).

The "conserved" variables are

Density

$$\overline{\rho}_k^x = \overline{X_k \rho} / \alpha_k \tag{14}$$

Velocity

$$\overline{\mathbf{v}}_k^{x\rho} = \overline{X_k \rho \mathbf{v}} / \alpha_k \overline{\rho}_k^x \tag{15}$$

The averaged stress is defined by

$$\overline{\mathbf{T}}_k^x = \overline{X_k \mathbf{T}} / \alpha_k \tag{16}$$

The stress at the interface gives rise to a source of momentum:

$$\mathbf{M}_k = -\overline{\mathbf{T} \cdot \nabla X_k} \tag{17}$$

The motion of the interfaces gives rise to velocities that are not "laminar" in general. The velocity fluctuations may be due to turbulence or to the motion in the phases due to the motion of the interfaces. The effect of these velocity fluctuations, whatever their source, on a variable is accounted for by introducing its fluctuating field (denoted by the prime superscript), which is the difference between the complete field and the appropriate mean field. For example,

$$\mathbf{v}_k' = \mathbf{v} - \overline{\mathbf{v}}_k^{x\rho}$$

Then

$$\begin{aligned} \overline{X_k \rho \mathbf{v}\mathbf{v}} &= \overline{X_k \rho (\overline{\mathbf{v}}_k^{x\rho} + \mathbf{v}_k')(\overline{\mathbf{v}}_k^{x\rho} + \mathbf{v}_k')} \\ &= \overline{X_k \rho} \overline{\mathbf{v}}_k^{x\rho} \overline{\mathbf{v}}_k^{x\rho} + \overline{X_k \rho \mathbf{v}_k' \mathbf{v}_k'} \\ &= \alpha_k \overline{\rho}_k^x \overline{\mathbf{v}}_k^{x\rho} \overline{\mathbf{v}}_k^{x\rho} - \alpha_k \mathbf{T}_k^{Re}. \end{aligned} \tag{18}$$

This defines the Reynolds stress

$$\mathbf{T}_k^{Re} = -\overline{X_k \rho \mathbf{v}_k' \mathbf{v}_k'}/\alpha_k \tag{19}$$

The averaged interfacial pressure p_{ki} and shear stress $\boldsymbol{\tau}_{ki}$ are introduced to separate mean field effects from local effects in the interfacial force.
Interfacial pressure

$$p_{ki} = \overline{p \partial X_k / \partial n_k}/a_i \tag{20}$$

Interfacial shear stress

$$\boldsymbol{\tau}_{ki} = \overline{\boldsymbol{\tau}_k \partial X_k / \partial n_k}/a_i \tag{21}$$

Thus,

$$\begin{aligned}
\mathbf{M}_k &= -\overline{\mathbf{T} \cdot \nabla X_k} \\
&= \overline{p \nabla X_k} - \overline{\boldsymbol{\tau} \cdot \nabla X_k} \\
&= p_{ki} \overline{\nabla X_k} - \boldsymbol{\tau}_{ki} \cdot \overline{\nabla X_k} - \overline{\mathbf{T}_{ki}' \cdot \nabla X_k} \\
&= p_{ki} \nabla \alpha_k - \boldsymbol{\tau}_{ki} \nabla \alpha_k + \mathbf{M}_k',
\end{aligned} \tag{22}$$

where we define the interfacial extra momentum source

$$\mathbf{M}_k' = \mathbf{M}_k + p_{ki} \nabla \alpha_k - \boldsymbol{\tau}_{ki} \cdot \nabla \alpha_k. \tag{23}$$

We now present the averaged equations. For simplicity, all notation for the averaging will be dropped. The averaged equations governing each phase are
Mass

$$\frac{\partial \alpha_k \rho_k}{\partial t} + \nabla \cdot \alpha_k \rho_k \mathbf{v}_k = \Gamma_k \tag{24}$$

Momentum

$$\frac{\partial \alpha_k \rho_k \mathbf{v}_k}{\partial t} + \nabla \cdot \alpha_k \rho_k \mathbf{v}_k \mathbf{v}_k = \nabla \cdot \alpha_k \left(\mathbf{T}_k + \mathbf{T}_k^{Re} \right) + \alpha_k \rho_k \mathbf{g} + \mathbf{M}_k + \mathbf{v}_{ki}^m \Gamma_k \tag{25}$$

4. Deriving Constitutive Equations. The equations of motion must be closed by supplying equations to relate the interactions of the materials on each other (interfacial terms) and the interaction with each material with itself. In the situations where the microscale problem can be solved the results obtained can be used to give information about the constitutive equations. We shall do this for the flow of an inviscid, incompressible irrotational fluid around an isolated sphere. We shall use these solutions to derive information about constitutive equations for the interfacial force, the average pressure, the Reynolds stress, and the interfacial pressure.

In irrotational flow, we have

$$\mathbf{v}(\mathbf{x}) = \nabla\phi(\mathbf{x}) \tag{26}$$

and the continuity equation becomes

$$0 = \nabla \cdot \mathbf{v} = \nabla \cdot \nabla\phi = \nabla^2\phi. \tag{27}$$

The pressure is given by Bernoulli's equation.

$$p = p_0 - \rho(\frac{\partial\phi}{\partial t} + \frac{1}{2}|\nabla\phi|^2) \tag{28}$$

where p_0 is a constant.

Consider a sphere located at a point $\mathbf{z}$ in a flow field, moving with velocity $\mathbf{v}_p(\mathbf{z}, t)$. The boundary condition at the surface of the sphere is

$$\mathbf{n} \cdot \mathbf{v}_s = \mathbf{n} \cdot \mathbf{v} = \mathbf{n} \cdot \nabla\phi \text{ at } |\mathbf{x} - \mathbf{z}| = a, \tag{29}$$

where a is the radius of the sphere, $\mathbf{n}$ is the normal to the surface of the sphere and $\mathbf{v}_p$ is the velocity of the sphere. The boundary condition far from the sphere is

$$\phi \to \phi_\infty \text{ as } |\mathbf{x} - \mathbf{z}| \to \infty, \tag{30}$$

where

$$\phi_\infty = \mathbf{v}_0(t) \cdot \mathbf{x} + \frac{1}{2}\mathbf{x} \cdot \mathbf{e}_f \cdot \mathbf{x}$$

is the velocity potential that would exist in the fluid if the sphere were not present. Here $\mathbf{v}_0(t)$ is the (unsteady) velocity of the fluid at the origin, and $\mathbf{e}_f$ is the rate of strain tensor for the fluid. We shall assume that $\mathbf{e}_f$ is constant.

A convenient form for the solution of this problem is given by Voinov (1973), and is

$$\begin{aligned}\phi =& \mathbf{v}_0(t) \cdot \mathbf{x} + \frac{1}{2}\mathbf{x} \cdot \mathbf{e}_f \cdot \mathbf{x} \\ &+ \frac{1}{2}(\mathbf{v}_p(\mathbf{z}, t) - \mathbf{v}_0(t) + \mathbf{z} \cdot \mathbf{e}_f) \cdot (\mathbf{x} - \mathbf{z}) \left(\frac{a^3}{r^3}\right) \\ &+ \frac{1}{3}(\mathbf{x} - \mathbf{z}) \cdot \mathbf{e}_f \cdot (\mathbf{x} - \mathbf{z}) \left(\frac{a^5}{r^5}\right).\end{aligned} \tag{31}$$

We assume that each sphere lies in a "cell," and inside that cell, the velocity is given by eq. (31). We shall approximate the cell to be a sphere of radius R. We choose R so that

$$\frac{4}{3}\pi a^3 / \frac{4}{3}\pi R^3 = \alpha_p.$$

We now introduce the averaging processes. Volume averaging essentially integrates expressions over large volumes, divides by the volume of the region, and assigns the average value to some central point of the averaging volume. For example, the averaging volume could be a sphere of radius L, with $L \gg R$. We assign averaged values of the flow quantities at different points in the flow field by moving the averaging volume around so that its center is at different points $\mathbf{z}$.

The cell model can be used to approximate the volume average. If the averaging volume is sufficiently large that there are many cells in the averaging volume, then it is customary to average over only one "representative" cell, and assign that average to the field point $\mathbf{z}$. There must be some decision made on the position of the cell relative to the field point $\mathbf{z}$. The practice seems to be that one performs the cell approximation to the volume average at some point $\mathbf{y}$ by assuming that $\mathbf{z} = \mathbf{y}$, and integrating over a volume in $\mathbf{x}$.

For the cell approximation to the ensemble average, the ensemble is the set of flows that can occur at $\mathbf{x}$ with the position of the sphere center occupying different positions in the cell. In this case, the average is performed by integrating over the possible positions that the sphere center can have. That is, the center of the sphere $\mathbf{z}$ can lie anywhere in the sphere of radius R. Note that if $|\mathbf{x}-\mathbf{z}| < a$, the material making up the sphere occupies the field point $\mathbf{x}$ and if $|\mathbf{x}-\mathbf{z}| > a$ the fluid occupies the field point $\mathbf{x}$.

These two averaging processes are different. The essential difference lies in the "background" motion to which the sphere responds.

The average velocity of the fluid phase is given by

$$\overline{\mathbf{v}}_f^{xp} = \frac{1}{\frac{4}{3}\pi(R^3 - a^3)} \int_a^R \int\int_{\Omega(r)} \mathbf{v}(\mathbf{x},\mathbf{z},t)\, d\Omega\, dr, \tag{32}$$

where $\Omega(r)$ is the sphere of radius r centered at $\mathbf{x}$, and the integration is over the $\mathbf{z}$ variable. We have

$$\begin{aligned}
\mathbf{v}(\mathbf{x},\mathbf{z},t) &= \nabla\phi(\mathbf{x},\mathbf{z},t) \\
&= \mathbf{v}_f(\mathbf{x}) + \frac{1}{2}(\mathbf{v}_f(\mathbf{z}) - \mathbf{v}_p(\mathbf{z}))\left(\frac{a^3}{r^3}\right) \\
&\quad - \frac{3}{2}(\mathbf{v}_f(\mathbf{z}) - \mathbf{v}_p(\mathbf{z}))\cdot(\mathbf{x}-\mathbf{z})\left(\frac{a^3}{r^5}\right)(\mathbf{x}-\mathbf{z}) \\
&\quad + \frac{2}{3}(\mathbf{x}-\mathbf{z})\cdot\mathbf{e}_f\left(\frac{a^5}{r^5}\right) - \frac{5}{3}(\mathbf{x}-\mathbf{z})\cdot\mathbf{e}_f\cdot(\mathbf{x}-\mathbf{z})\left(\frac{a^5}{r^7}\right)(\mathbf{x}-\mathbf{z}).
\end{aligned} \tag{33}$$

Note that $\mathbf{v}_f(\mathbf{x})$ is the fluid velocity that would exist at $\mathbf{x}$ if the sphere were not present, and $\mathbf{v}_f(\mathbf{z}) - \mathbf{v}_p(\mathbf{z})$ is the relative velocity between the sphere and the fluid evaluated at the sphere center. In order to evaluate the integrals appearing in eq. (31), we must express the $\mathbf{z}$ dependence of the velocities in terms of $\mathbf{x}$ and $\mathbf{x}' = \mathbf{x}-\mathbf{z}$. We have

$$\mathbf{v}_f(\mathbf{z}) = \mathbf{v}_f(\mathbf{x}) - \mathbf{x}'\cdot\mathbf{e}_f$$

and

$$\mathbf{v}_p(\mathbf{z}) = \mathbf{v}_p(\mathbf{x}) - \mathbf{x}' \cdot \mathbf{e}_p$$

where $\mathbf{e}_p$ is the velocity gradient tensor for the particle motion. We shall assume that this tensor is constant and symmetric. Substituting this into eq. (8) gives

$$\begin{aligned} \mathbf{v}(\mathbf{x}, \mathbf{z}, t) == &\mathbf{v}_f(\mathbf{x}) + \frac{1}{2}(\mathbf{v}_f(\mathbf{x}) - \mathbf{v}_p(\mathbf{x}) - \mathbf{x}' \cdot (\mathbf{e}_f - \mathbf{e}_p)) \left(\frac{a^3}{r^3}\right) \\ &- \frac{3}{2}(\mathbf{v}_f(\mathbf{x}) - \mathbf{v}_p(\mathbf{x}) - \mathbf{x}' \cdot (\mathbf{e}_f - \mathbf{e}_p)) \cdot \mathbf{x}' \left(\frac{a^3}{r^5}\right) \mathbf{x}' \\ &+ \frac{2}{3}\mathbf{x}' \cdot \mathbf{e}_f \left(\frac{a^5}{r^5}\right) - \frac{5}{3}\mathbf{x}' \cdot \mathbf{e}_f \cdot \mathbf{x}' \left(\frac{a^5}{r^7}\right) \mathbf{x}'. \end{aligned} \tag{34}$$

It is convenient to have expressions for the integrals of powers of $\mathbf{x}'$ over $\Omega(r)$. For these integrals, we note that

$$\int_{\Omega(r)} \mathbf{x}' \ldots \mathbf{x}' d\Omega = 0 \tag{35}$$

if the factor $\mathbf{x}'$ appears on odd number of times, and

$$\int_{\Omega(r)} d\Omega = 4\pi r^2 \tag{36a}$$

$$\int_{\Omega(r)} \mathbf{x}'\mathbf{x}' d\Omega = \frac{4}{3}\pi r^4 \mathbf{I} \tag{36b}$$

$$\int_{\Omega(r)} \mathbf{x}'\mathbf{x}'\mathbf{x}'\mathbf{x}' d\Omega = \frac{4}{15}\pi r^6 \mathbf{\Sigma} \tag{36c}$$

where $\mathbf{\Sigma}$ is a fourth order isotropic tensor defined in Cartesian coordinates by

$$\Sigma_{ijkl} = \delta_{ij}\delta_{kl} + \delta_{ik}\delta_{jl} + \delta_{il}\delta_{jk}.$$

We further note that if $\mathbf{v}$ is a vector, and $\mathbf{e}$ is a symmetric second order tensor with $e_{ii} = 0$, then

$$\Sigma_{ijkl} v_j e_{kl} = 2 v_j e_{ji}.$$

Using these results in eq. (32) gives

$$\overline{\mathbf{v}}_f^{xp} = \mathbf{v}_f(\mathbf{x}, t). \tag{37}$$

The interfacial averaged velocity of the fluid is given by

$$\overline{\mathbf{v}}_{ci} = \frac{1}{4\pi a^2} \int_{\Omega(a)} \mathbf{v}(\mathbf{x}, \mathbf{z}, t) d\Omega. \tag{38}$$

Substituting and performing the integrations lead to the result

$$\overline{\mathbf{v}}_{ci} = \mathbf{v}_f(\mathbf{x},t). \tag{39}$$

This result is a little surprising at first. The fluid at the surface of the sphere satisfies the condition $\mathbf{n}\cdot\mathbf{v} = \mathbf{n}\cdot\mathbf{v}_p$, but is allowed to slip in the tangential direction. After the passage of the sphere, the fluid that was momentarily in contact with the surface of the sphere is again moving with the fluid. The result says that even during the time that it is in contact with the surface of the sphere, its average velocity is still equal to the average velocity of the fluid, and not of the sphere.

Now let us compute averaged pressures using this formalism. The exact pressure can be computed by Bernoulli's equation (28)

$$p = p_0 - \rho_c\left(\frac{1}{2}|\nabla\phi|^2 + \frac{\partial\phi}{\partial t}\right). \tag{40}$$

In order to evaluate the derivatives in eq. (40), we note that $\mathbf{x}$ is constant during t derivatives, but $\partial\mathbf{z}/\partial t = \mathbf{v}_p(\mathbf{z})$. Also, when evaluating $\nabla\phi$, both t and $\mathbf{z}$ are held constant. The pressure is given by

$$\begin{aligned}
p = p_0 - \rho_c\Bigg(&\frac{\partial\mathbf{v}_0}{\partial t}\cdot\mathbf{x} + \frac{1}{2}\left[\frac{\partial\mathbf{v}_0}{\partial t} - \frac{\partial\mathbf{v}_p}{\partial t} + \mathbf{v}_p(\mathbf{z})\cdot\mathbf{e}_f - \mathbf{v}_p(\mathbf{z})\cdot\mathbf{e}_p\right]\cdot\mathbf{x}'\left(\frac{a^3}{r^3}\right)\\
&-\frac{1}{2}(\mathbf{v}_f(\mathbf{x}) - \mathbf{v}_p(\mathbf{x}) - \mathbf{x}'\cdot(\mathbf{e}_f - \mathbf{e}_p))\cdot\mathbf{v}_p(\mathbf{z})\left(\frac{a^3}{r^3}\right)\\
&-\frac{3}{2}(\mathbf{v}_f(\mathbf{x}) - \mathbf{v}_p(\mathbf{x}) - \mathbf{x}'\cdot(\mathbf{e}_f - \mathbf{e}_p))\cdot\mathbf{x}'\mathbf{v}_p(\mathbf{z})\left(\frac{a^3}{r^5}\right)\mathbf{x}'\cdot\mathbf{v}_p(\mathbf{z})\\
&-\frac{2}{3}\mathbf{v}_p(\mathbf{z})\cdot\mathbf{e}_f\cdot\mathbf{x}'\left(\frac{a^5}{r^5}\right) + \frac{5}{3}\mathbf{x}'\cdot\mathbf{e}_f\cdot\mathbf{x}'\mathbf{x}'\cdot\mathbf{v}_p(\mathbf{z})\left(\frac{a^5}{r^5}\right) + \frac{1}{2}\mathbf{v}_f\cdot\mathbf{v}_f\\
&+\frac{1}{8}(\mathbf{v}_f(\mathbf{x}) - \mathbf{v}_p(\mathbf{x}) - \mathbf{x}'\cdot(\mathbf{e}_f - \mathbf{e}_p))\cdot(\mathbf{v}_f(\mathbf{x}) - \mathbf{v}_p(\mathbf{x}) - \mathbf{x}'\cdot(\mathbf{e}_f - \mathbf{e}_p))\left(\frac{a^6}{r^6}\right)\\
&+\frac{1}{2}\mathbf{v}_f\cdot(\mathbf{v}_f(\mathbf{x}) - \mathbf{v}_p(\mathbf{x}) - \mathbf{x}'\cdot(\mathbf{e}_f - \mathbf{e}_p))\left(\frac{a^3}{r^3}\right)\\
&+\frac{9}{8}\left((\mathbf{v}_f(\mathbf{x}) - \mathbf{v}_p(\mathbf{x}) - \mathbf{x}'\cdot(\mathbf{e}_f - \mathbf{e}_p))\cdot\mathbf{x}'\right)^2\left(\frac{a^6}{r^{10}}\right)\\
&+\frac{2}{3}\mathbf{v}_p(\mathbf{x})\cdot\mathbf{e}_f\cdot\mathbf{x}'\left(\frac{a^5}{r^5}\right) - \frac{5}{3}\mathbf{v}_p(\mathbf{x})\cdot\mathbf{x}'\mathbf{x}'\cdot\mathbf{e}_f\cdot\mathbf{x}'\left(\frac{a^5}{r^7}\right)\\
&\frac{1}{3}(\mathbf{v}_f(\mathbf{x}) - \mathbf{v}_p(\mathbf{x}))\cdot\mathbf{e}_f\cdot\mathbf{x}'\left(\frac{a^8}{r^8}\right) - \frac{5}{6}(\mathbf{v}_f(\mathbf{x}) - \mathbf{v}_p(\mathbf{x}))\cdot\mathbf{x}'\mathbf{x}'\cdot\mathbf{e}_f\cdot\mathbf{x}'\left(\frac{a^8}{r^{10}}\right)\\
&+\frac{3}{2}(\mathbf{v}_f(\mathbf{x}) - \mathbf{v}_p(\mathbf{x}))\cdot\mathbf{x}'\mathbf{x}'\cdot\mathbf{e}_f\cdot\mathbf{x}'\left(\frac{a^8}{r^{10}}\right)\Bigg)
\end{aligned} \tag{41}$$

where we have ignored terms of order e_f^2, e_p^2, and $e_f e_p$.

The calculations are tedious, but result in

$$\overline{p}_c^x = p_0 - \rho_c \frac{\partial \mathbf{v}_f}{\partial t} \cdot \mathbf{x} - \frac{1}{2}\mathbf{v}_f(\mathbf{x}) \cdot \mathbf{v}_f(\mathbf{x}) - \frac{1}{4}\alpha_d \rho_c |\mathbf{v}_f(\mathbf{x}) - \mathbf{v}_p(\mathbf{x})|^2 \tag{42}$$

$$\overline{p}_{ci} = \overline{p}_c^x - \frac{1}{4}\rho_c |\mathbf{v}_c^{xp} - \mathbf{v}_d^{xp}|^2. \tag{43}$$

where we ignore terms of order α_d^2 in addition to those ignored previously.

The interfacial force density $\mathbf{M}_d$ is given by $\mathbf{M}_d = \overline{p\nabla X_d}$. Thus,

$$\mathbf{M}_d = -\frac{1}{\frac{4}{3}\pi R^3} \int_{\Omega(a)} \mathbf{n} p(\mathbf{x}, \mathbf{z}, t) d\Omega. \tag{44}$$

This can be computed by substituting eq. (28) for the pressure, and recognizing that $\mathbf{n} = \mathbf{x}'/a$. The result is that

$$\begin{aligned} \mathbf{M}_d = \alpha_d \rho_c & \left(\frac{1}{2} \left[\frac{\partial \mathbf{v}_0}{\partial t} - \frac{\partial \mathbf{v}_p}{\partial t} + \mathbf{v}_f \cdot \mathbf{e}_f - \mathbf{v}_p \cdot \mathbf{e}_p \right] \right. \\ & \left. - \frac{7}{20}(\mathbf{v}_f - \mathbf{v}_p) \cdot (\mathbf{e}_f - \mathbf{e}_p) \right). \end{aligned} \tag{45}$$

Note that no drag force is present in eq. (45). This is the result of D'Alembert's paradox, that is, there is no net force on a body moving at a constant velocity through an inviscid fluid at rest.

It is also possible to calculate the force on a sphere at $\mathbf{z}$ by computing

$$\mathbf{F}_p(\mathbf{z}) = \int_{\Omega(a)} \mathbf{n} \left(p_0 - \rho_c \left[\frac{1}{2}|\nabla \phi|^2 + \frac{\partial \phi}{\partial t} \right] \right) d\Omega,$$

where the integration is over the variable $\mathbf{x}'$, with $\mathbf{x} = \mathbf{z} + \mathbf{x}'$. This results in

$$\mathbf{F}_p(\mathbf{z}) = \frac{4}{3}\pi a^3 \rho_c \left(\frac{\partial \mathbf{v}_f}{\partial t} + \mathbf{v}_f \cdot \mathbf{e}_f + \frac{1}{2} \left[\frac{\partial \mathbf{v}_f}{\partial t} - \frac{\partial \mathbf{v}_p}{\partial t} + \mathbf{v}_f \cdot \mathbf{e}_f \right] \right). \tag{46}$$

Note that this force agrees with Taylor's calculation of the force necessary to hold a sphere at rest in an accelerating stream, obtained by setting $\frac{\partial}{\partial t} = 0$ and $\mathbf{v}_p = 0$.

In order to illustrate the difference between the ensemble average and the volume average, let us compute the volume average of $\mathbf{M}_d^V = \overline{p\nabla X_d}$. We have

$$\mathbf{M}_d^V = -\frac{1}{\frac{4}{3}\pi R^3} \int_{\Omega(a)} \mathbf{n} p(\mathbf{x}, \mathbf{z}, t) d\Omega, \tag{47}$$

where now the integration is over $\mathbf{x} = \mathbf{z} + \mathbf{x}'$ keeping $\mathbf{z}$ fixed. The result is

$$\mathbf{M}_d^V = \alpha_d \rho_c \left(\frac{\partial \mathbf{v}_0}{\partial t} + \mathbf{v}_f \cdot \mathbf{e}_f + \frac{1}{2}\alpha_d \rho_c \left[\frac{\partial \mathbf{v}_0}{\partial t} - \frac{\partial \mathbf{v}_p}{\partial t} + \mathbf{v}_f \cdot \mathbf{e}_f \right] \right). \tag{45}$$

It is interesting to note that this term, BY ITSELF, gives Taylor's result for the force on a sphere in a spatially accelerating flow. This means that the result will be incorrect if used in an equation where a force due to the fluid pressure gradient is also present. The ultimate result will include the fluid acceleration TWICE. A similar thing happens if we include the potential due to the gravity head. Using the ensemble average, no gravitational force appears in the interfacial force. However, if we use the volume average, the buoyancy term $\alpha_d \rho_c \mathbf{g}$ appears in the interfacial force. Then the force due to the fluid pressure gradient will also include the same force, and hence it, too will appear twice.

We next turn to computations of the Reynolds stress using the velocity fluctuations due to the inviscid flow around a sphere. Using the expression for the velocity (9), we see that

$$\begin{aligned} \mathbf{v}'_c(\mathbf{x},\mathbf{z},t) &= \frac{1}{2}(\mathbf{v}_f(\mathbf{z}) - \mathbf{v}_p(\mathbf{z}))\left(\frac{a^3}{r^3}\right) \\ &- \frac{3}{2}(\mathbf{v}_f(\mathbf{z}) - \mathbf{v}_p(\mathbf{z}))\cdot(\mathbf{x}-\mathbf{z})\left(\frac{a^3}{r^5}\right)(\mathbf{x}-\mathbf{z}) \\ &+ \frac{2}{3}(\mathbf{x}-\mathbf{z})\cdot\mathbf{e}_f\left(\frac{a^5}{r^5}\right) - \frac{5}{3}(\mathbf{x}-\mathbf{z})\cdot\mathbf{e}_f\cdot(\mathbf{x}-\mathbf{z})\left(\frac{a^5}{r^7}\right)(\mathbf{x}-\mathbf{z}). \end{aligned} \tag{46}$$

Therefore $\mathbf{T}_c^{Re} = -\rho_c \overline{\mathbf{v}'\mathbf{v}'}^{xp}$ can be computed, and the result is

$$\mathbf{T}_c^{Re} = -\frac{1}{20}\alpha_d \overline{\rho}_c\left((\overline{\mathbf{v}}_c^{xp} - \overline{\mathbf{v}}_d^{xp})(\overline{\mathbf{v}}_c^{xp} - \overline{\mathbf{v}}_d^{xp}) + 3(\overline{\mathbf{v}}_c^{xp} - \overline{\mathbf{v}}_d^{xp})\cdot(\overline{\mathbf{v}}_c^{xp} - \overline{\mathbf{v}}_d^{xp})\mathbf{I}\right). \tag{47}$$

The fluid fluctuation kinetic energy is $u_c^{Re} = \frac{1}{2}\overline{\mathbf{v}'\cdot\mathbf{v}'}^{x}$, and can be computed by taking the trace of eq. (47) for $\mathbf{T}_c^{Re}$. The result is

$$u_c^{Re} = \frac{1}{4}\alpha_d(\overline{\mathbf{v}}_c^{xp} - \overline{\mathbf{v}}_d^{xp})\cdot(\overline{\mathbf{v}}_c^{xp} - \overline{\mathbf{v}}_d^{xp}). \tag{48}$$

The stress inside the spheres is another matter. It is clear that the stresses around a sphere are transmitted somehow through the sphere. We shall assume that the sphere is a linearly elastic solid, but we shall assume that the strain is sufficiently small that the deformation is undetectable from the outside. Then the stress-strain relation is given by

$$\mathbf{T} = \mu[\nabla\mathbf{u} + (\nabla\mathbf{u})^{tr}] + \lambda\nabla\cdot\mathbf{u}\mathbf{I} \tag{49}$$

This can be written as

$$\mathbf{T} = \mu[\nabla\mathbf{u} + (\nabla\mathbf{u})^{tr} - \frac{2}{3}\nabla\cdot\mathbf{u}\mathbf{I}] + \Theta\mathbf{I} \tag{50}$$

The spherical part of the stress satisfies (Love, 1927)

$$\nabla^2\Theta = 0$$

$$\Theta = -p \text{ on } |\mathbf{x} - \mathbf{z}| = a \tag{51}$$

The computation gives

$$\overline{\Theta}_d^x(\mathbf{x}, t) = -p_i(\mathbf{x}, t) \tag{52}$$

If we then consider the rest of the stress tensor, given by

$$\sigma = \mu[\nabla \mathbf{u} + (\nabla \mathbf{u})^{tr} - \frac{2}{3}\nabla \cdot \mathbf{u}\mathbf{I}]$$

the solution is again given in Love (1927) and can be averaged to give

$$\overline{\sigma}_d^x(\mathbf{x}, t) = -\rho_c \left(-\frac{9}{20}(\mathbf{v}_f - \mathbf{v}_p)(\mathbf{v}_f - \mathbf{v}_p) - \frac{3}{20}|(\mathbf{v}_f - \mathbf{v}_p)|^2 \mathbf{I} \right) \tag{53}$$

Substituting eqs. (45), (52), (53) into eq. (25) for the particle phase gives

$$\begin{aligned}
\alpha_d \rho_d \left(\frac{\partial \mathbf{v}_d}{\partial t} + \mathbf{v}_d \cdot \nabla \mathbf{v}_d \right) &= -\alpha_d \nabla p_c \\
+\frac{1}{2}\alpha_d \rho_c \left[\left(\frac{\partial \mathbf{v}_c}{\partial t} + \mathbf{v}_c \cdot \nabla \mathbf{v}_c \right) - \left(\frac{\partial \mathbf{v}_d}{\partial t} + \mathbf{v}_d \cdot \nabla \mathbf{v}_d \right) \right] & \\
&+\frac{1}{2}\alpha_d \rho_c [\nabla(\mathbf{v}_c - \mathbf{v}_d)] \cdot (\mathbf{v}_c - \mathbf{v}_d) \\
&-\frac{7}{20}\alpha_d \rho_c (\mathbf{v}_c - \mathbf{v}_d) \cdot (\mathbf{v}_c - \mathbf{v}_d) \\
&-\frac{9}{20}\alpha_d \rho_c (\mathbf{v}_c - \mathbf{v}_d) \cdot (\mathbf{v}_c - \mathbf{v}_d) \\
&+\frac{6}{20}\alpha_d \rho_c [\nabla(\mathbf{v}_c - \mathbf{v}_d)] \cdot (\mathbf{v}_c - \mathbf{v}_d)
\end{aligned} \tag{54}$$

This result agrees with Taylor's for the force on a single sphere in an accelerating flow. To see this, let $\mathbf{v}_d = 0$, and $\partial/\partial t = 0$. For small particle concentration, the fludi momentum equation reduces to $-\nabla p_c = \rho_c \mathbf{v}_c \cdot \nabla \mathbf{v}_c$. Then the right hand side of (54) reduces to $\frac{3}{2}\alpha_d \rho_c \mathbf{v}_c \cdot \mathbf{v}_c$.

Acknowledgements. This work was supported in part by the Institute for Mathematics and its Applications, Minneapolis, Minnesota and by the Mathematical Sciences Institute, Cornell University.

REFERENCES

[1] J. H. STUHMILLER, *The influence of interfacial pressure on the character of two-phase flow model equations International Journal of Multiphase Flow.*

[2] LOVE, A. E. H., *A Treatise on the Mathematical Theory of Elasticity*, Cambridge University Press, London, 1927.

THE VELOCITY OF DYNAMIC WAVES IN FLUIDISED BEDS

LARRY GIBILARO*, PIERO FOSCOLO† AND RENZO DI FELICE†

Abstract. A simple expression for the velocity of dynamic waves in fluidised suspensions that has been derived solely in terms of the equilibrium fluid- particle interaction force parameters, is shown to be in close agreement with direct measurements made under incipient fluidisation conditions; further support is provided by the extensive minimum bubbling point data reported for both gas and liquid fluidised systems and other reported observations of fluidisation quality.

Key words. fluidisation, particle pressure.

1. Introduction. A powder contained in a vertical tube on a porous support and subjected to an upflow of fluid experiences an upwards acting force that increases with increasing fluid velocity. At the fluid velocity where this force just balances the weight of the powder, the system is said to become incipiently fluidised: effectively no particle mobility is discernible at this stage, except for some local short-lived rearrangements involving disruption of particle bridges and other inhomogeneities associated with the initial introduction of the powder to the tube. The subsequent behaviour of the bed as the fluid velocity is increased beyond its minimum fluidisation value depends critically on the properties of both the fluid and the particles employed. If, for example, 0.3mm glass particles are fluidised with water, the bed expands maintaining an essentially homogeneous suspension that simply becomes progressively less concentrated as the terminal condition, that corresponds to a single particle supported in the fluid stream, is approached; the bed surface remains quite flat for all expansion conditions; this is the behaviour normally associated with liquid fluidisation in general. If the same particles are now fluidised by air, the situation is quite different: all air, in excess of that required to bring the bed to the incipient fluidisation condition, forms small bubbles close to the porous base that travel up the bed faster than the gas permeating through the particles; these bubbles grow rapidly, mainly by coalescence, to erupt through the bed surface giving it the appearance of a vigorously boiling liquid; this is the behaviour normally associated with gas fluidisation; in contrast to normal liquid fluidisation, the state of essentially homogeneous suspension is unattainable in practice.

Exceptions to these general classifications are commonplace and occupy a prime focus in what follows: beds of finer powders when fluidised by a gas can exhibit an initial period of homogeneous expansion followed, after a critical gas velocity is reached, by the bubbling regime; this transition can be quite dramatic with the bed surface passing from a completely flat stable condition, through a violently oscillating state at the critical point, to the bubbling condition normally associated

*Department of Chemical and Biochemical Engineering, University College London, Torrington Place, London WC1E 7JE, U.K.

†Dipartimento di Chimica, Ingegneria Chimica e Materiale, Universita di L'Aquila, 67100, L'Aquila, ITALY.

with gas fluidisation, all within a very small range of variation of the fluid velocity. Somewhat similar behaviour can be observed when high density solids (for example, lead or copper spheres) are fluidised with water: a well-defined transfer from stable to bubbling behaviour occurs at a critical liquid velocity, although this time the violent oscillations at the transition point are absent. Lower density solids, such as glass spheres, can also switch from stable to bubbling behaviour when fluidised by water, but in this case the transition is a diffuse one characterised by a significant fluid velocity range in which horizontal high voidage bands can be observed travelling up the bed from the distributor region in a regular periodic manner; 'bubbles' that can be formed in these systems are disc shaped rather than near spherical and travel up the bed relatively slowly.

The purpose of this paper is to show that the characteristics of fluidisation quality outlined above are predictable in a quantitative sense on the basis of a remarkably simple formulation of the equations describing conservation of mass and momentum for the particle phase.

2. The Particle Bed Model. We are concerned with the one-dimensional formulation of conservation equations for mass and momentum of the particle phase [8]:

(2.1)
$$\frac{\partial \epsilon}{\partial t} - \frac{\partial}{\partial z}\left[(1-\epsilon)v\right] = 0,$$
$$(1-\epsilon)\rho_p\left[\frac{\partial v}{\partial t} + \frac{v\partial v}{\partial z}\right] = F(\epsilon, v) + u_e^2\rho_p\frac{\partial \epsilon}{\partial z}.$$

The terms on the right hand side of the momentum equation may be expressed as explicit functions of the basic properties of the particles, ρ_p and d_p, and the fluid, ρ_f and μ:

(2.2)
$$F = (\rho_p - \rho_f)g\left[\left(\frac{u_0 - v}{u_t}\right)^{\frac{4.8}{n}}\epsilon^{-4.8} - 1\right]\epsilon(1-\epsilon),$$

(2.3)
$$u_e^2 = \frac{2d_p}{3\rho_p}\frac{\partial F}{\partial \epsilon} = 3.2gd_p(1-\epsilon_0)(\rho_p - \rho_f)/\rho_p.$$

These expressions relate to perturbations to an equilibrium bed (of void fraction ϵ_0 fluidised by a volumetric flux of fluid u_0) for which the equilibrium relationship is given by:

(2.4)
$$u_0 = u_t\,\epsilon_0^n\,,$$

u_t being the terminal settling velocity of a single unhindered particle in the stagnant fluid. Correlations for u_t and the empirical exponent, n, are readily available in terms of the basic fluid and particle properties [4,2].

The form for F has been obtained by deducing from the equilibrium relationship, eqn. (2.4), expressions for the drag and pressure gradient (buoyancy) contribution that together support the particle weight, and applying the same functional dependency on ϵ and v to the general non-equilibrium state [7,8]. The expression for the dynamic wave velocity, u_e, comes about by considering the net effect of the forces (regarded solely as functions of particle concentration, or void fraction, ϵ) acting at the two horizontal surfaces of a control volume when a voidage gradient is imposed on a system initially at equilibrium. For a particle system containing many particles, a surface corresponds to a single particle layer: the ratio of particles per unit area in a layer to particles per unit volume is $2d_p/3$ so that the net force becomes:

$$\frac{2}{3}d_p\,\frac{\partial F}{\partial \epsilon}\,\frac{\partial \epsilon}{\partial z},$$

leading to eqn. (2.3).

The main purpose of this paper is to gather together the large body of data that provide direct and indirect support for the dynamic wave velocity expression: we turn first to the direct evidence.

3. Experimental Determination of the Dynamic Wave Velocity, u_e. Direct measurements of dynamic waves in fluidised beds are not generally possible as they either grow into shocks or decay away according to the particular system properties and conditions of operation. Wallis [19] however has described a method for measuring their velocity at incipient fluidisation conditions, and results using his technique have been obtained for cases of water fluidisation [13].

The method involves fluidising particles in a bed fitted with a mesh screen at the top. By first increasing the fluid velocity, particles become packed against this screen; by then reducing the flow to a value between one and two times the minimum fluidisation velocity, particles at the bottom interface of the packed section 'rain down' giving rise to an easily measurable dynamic shock wave.

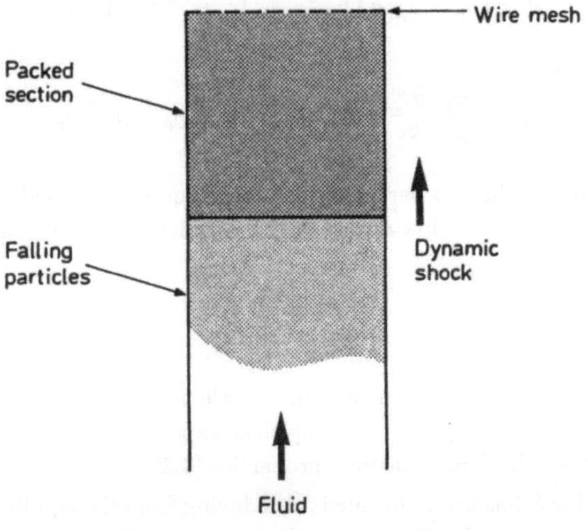

Fig. 3.1: Measurement of dynamic wave velocities at incipient fluidisation conditions.

By performing this experiment at a number of fluid velocities, the dynamic wave velocity at minimum fluidisation conditions can be obtained by extrapolation. Fig. 3.2 gives the results of dynamic shock velocities reported in [13] normalised with respect to the dynamic wave velocity expression of eqn. (2.3): convergence to the predicted value at u_{mf} is clearly indicated by these results and shown specifically in Fig. 3.3. Further confirmation is provided by the results obtained by Wallis [19] which are reproduced in Fig. 3.4 normalised in the same way; most of these results were obtained for air fluidisation.

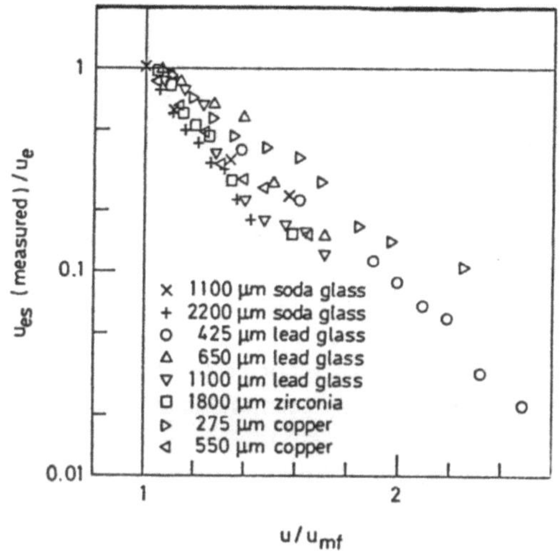

Fig. 3.2: Dynamic shock measurements in water fluidised beds normalised with respect to dynamic wave velocity predictions at $\epsilon_0 = 0.4$.

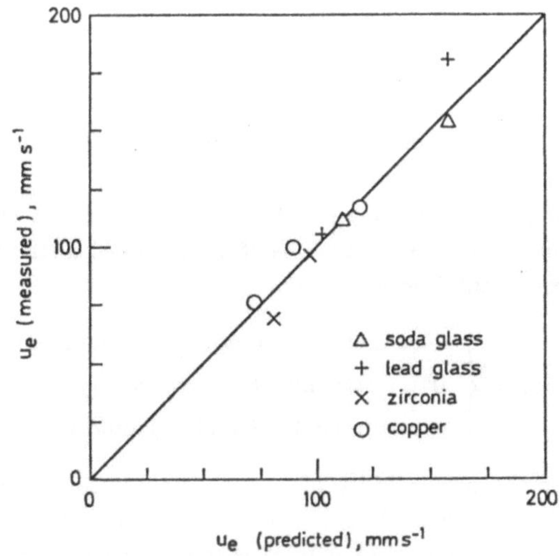

Fig. 3.3 Comparison of measured dynamic wave velocities with model predictions for water fluidised beds.

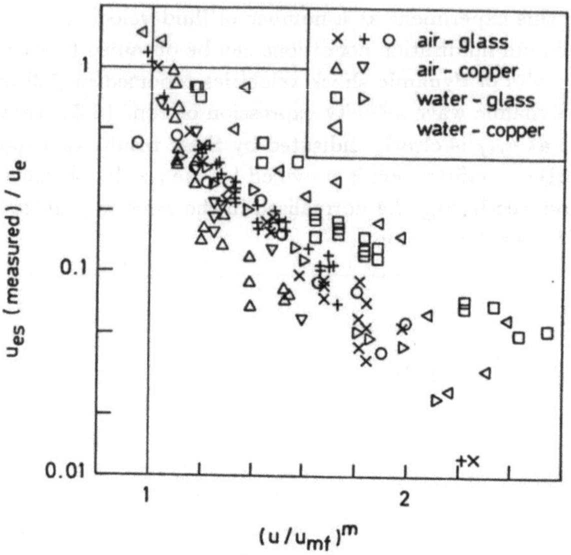

Fig. 3.4: Dynamic shock measurements reported by Wallis [19] normalised with respect to dynamic wave velocity predictions at $\epsilon_0 = 0.4$.

4. Wallis's Criterion for Instability. Linearisation of eqn. (2.1) about the equilibrium state, $v = 0$ and $\epsilon = \epsilon_0$, gives rise to the general one-dimensional form considered by Wallis [19,20] for small perturbations in void fraction:

$$\frac{\partial^2 \epsilon}{\partial t^2} - u_e^2 \frac{\partial^2 \epsilon}{\partial z^2} + B \left[\frac{\partial \epsilon}{\partial t} + u_\epsilon \frac{\partial \epsilon}{\partial z} \right] = 0, \tag{4.1}$$

$$u_\epsilon = u_t \, n(1 - \epsilon_0) \, \epsilon_0^{n-1},$$

$$B = 4.8g \, (\rho_p - \rho_f)/(n \, u_t \, \rho_p \, \epsilon_0^{n-1}).$$

The expression for u_ϵ represents the well-established form for the continuity wave velocity in a fluidised bed [18].

Wallis's criterion follows from eqn. (4.1) and, on the basis of the explicit expressions for u_e and u_ϵ, enables predictions of fluidised bed instability to be readily compared with experimental observations:

$$u_e - u_\epsilon = \begin{cases} +ve & : \text{stable (homogeneous fluidisation)} \\ 0 & : \text{stability limit } (\epsilon_0 = \epsilon_{mb}) \\ -ve & : \text{unstable (bubbling fluidisation)} \end{cases} \tag{4.2}$$

Fig. 4.1 illustrates the applicability of the criterion to four examples for which dynamic and continuity wave velocities have been evaluated from the expressions given above with values of u_t and n from standard correlations. The first case represents a typical bubbling gas fluidised bed and the second a typically stable liquid fluidised system; the other two cases show a transition from the stable to the bubbling state at a critical fluid velocity in broad agreement, as we report below, with experimental observations.

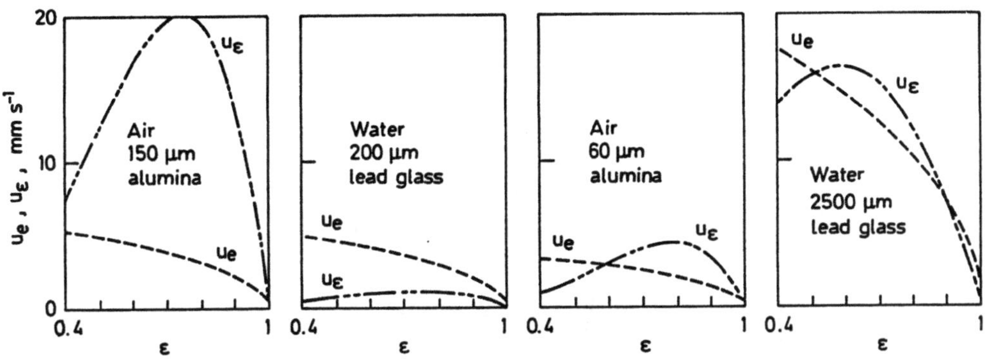

Fig. 4.1: Predicted continuity and dynamic wave velocities for various fluidised systems.

5. Global Predictions of Instability for Fluidisation with a Particular Fluid. The criterion for instability can be used to evaluate the combinations of particle density and size that, for a specified fluid, define the boundaries of completely stable and completely bubbling behaviour; the region between these boundaries represents particles that are predicted to transfer from a stable to a bubbling state at a critical fluid velocity. Fig. 5.1 illustrates this for air and water fluidisation: from a practical point of view it is the right hand boundary for air, indicating exceptions to the 'normal' bubbling state, and the left hand boundary for water, that are of most significance.

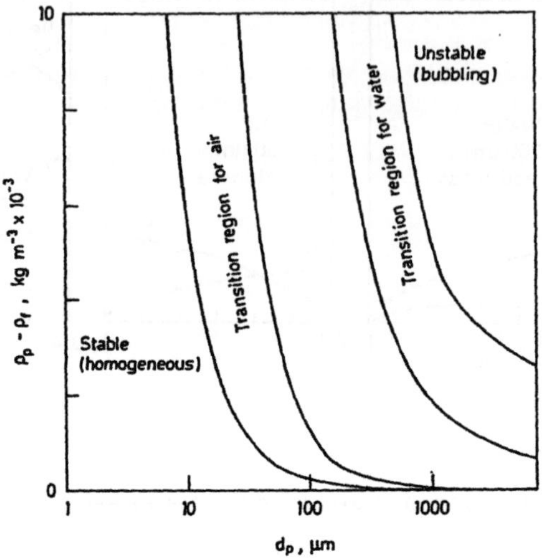

Fig. 5.1: Global predictions of bed stability for fluidisation by ambient air and water.

A substantial quantity of empirical data for the bubbling to transition region boundary for ambient air fluidisation has been correlated by Geldart [9]: his correlation is compared with predictions of the instability criterion in Fig. 5.2.

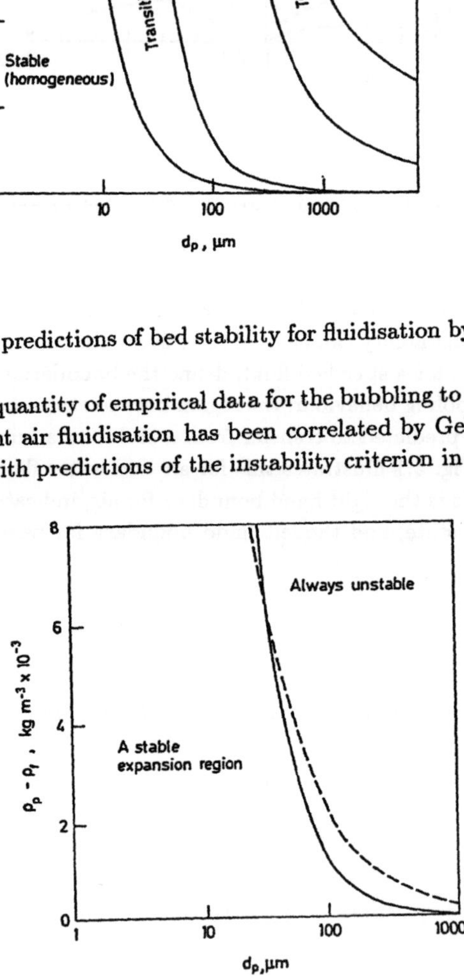

Fig. 5.2: Comparison of global predictions of a stable expansion region for ambient air fluidisation (continuous line) with empirical correlation of Geldart [9] (broken line).

6. The Minimum Bubbling Point for Gas Fluidisation of Fine Powders. Direct data on the transition from stable to bubbling behaviour of gas fluidised fine powders have been reported by many workers and collated elsewhere [8,

12]; they include results for air at other than ambient conditions and gases other than air; comparisons with the predictions of eqn. (4.2) are reproduced in Fig. 6.1.

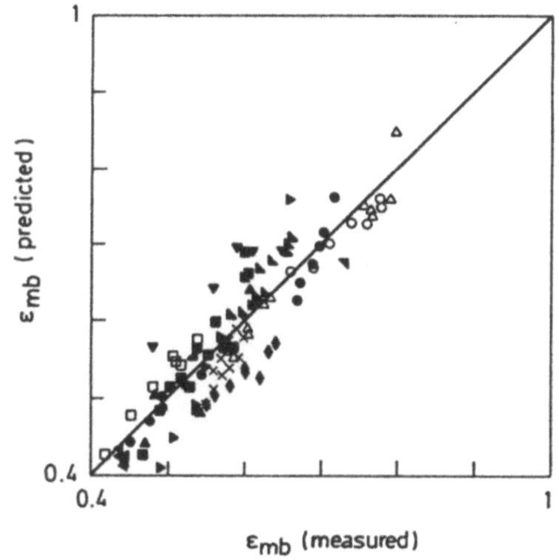

Fig. 6.1: Comparison of minimum bubbling point data for gas fluidisation reported by various authors with model predictions. (Source, Foscolo and Gibilaro [8] and Gibilaro et al. [12]).

Further support for the predictive ability of eqn. (4.2) is provided by experiments in which the trend in the minimum bubbling voidage, ϵ_{mb}, with systematic variation of a single system property is reported. This is illustrated in Fig. 6.2 for every variable that enters into the formulation - including the effect of gravitational field strength variation, which was simulated using a large-scale centrifuge [16].

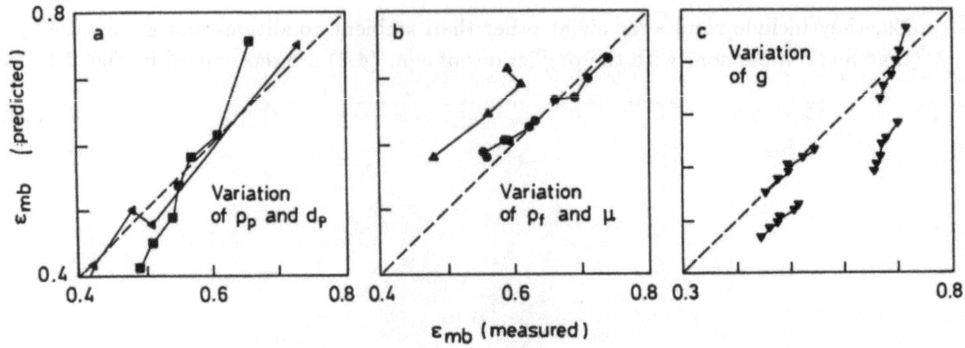

Fig. 6.2 : The effect on the minimum bubbling point of systematic variation of each hydrodynamic parameter (ρ_p, d_p, ρ_f, μ, g): comparison with model predictions.

a) Variation of particle properties: particle size (squares), De Jong and Nomden [5]; particle density (triangles), various authors summarised in Gibilaro et al, [12].

b) Variation of fluid properties: fluid density (circles), Jacob and Weimer [15] and Crowther and Whitehead [13]; fluid viscosity (triangles), Rowe [17].

c) Variation of effective gravitational field strength: Rietema and Mutsers [16].

7. Liquid Fluidisation. Although the phenomenon of bubbling in liquid fluidised beds has been known about for a long time [21], few systematic data have been reported until recently : Fig. 7.1 shows the global predictions of relation (4.2) for fluidisation by ambient water together with the key results of an experimental investigation into bubbling liquid fluidisation [10].

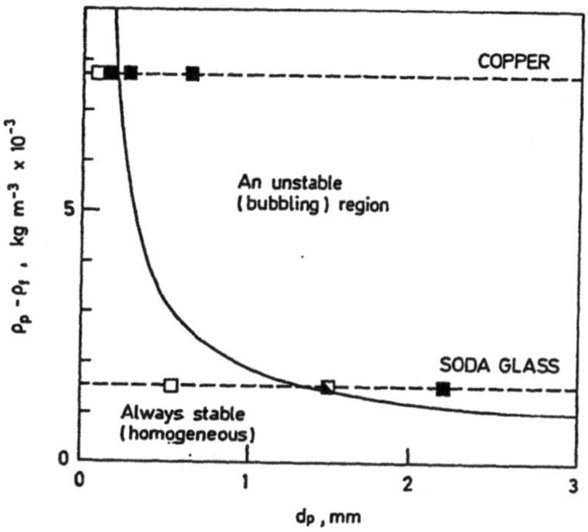

Fig. 7.1: Global predictions for some unstable (bubbling) behaviour for fluidisation by ambient water.

Points show experimental observations (Gibilaro et al, [10]): open points, complete stability; solid points, some bubbling; half solid, 'indeterminate stability.'

The observed behaviour of these systems has been referred to in section 1 and it can be seen that there is a fair measure of quantitative agreement with the global predictions: in fact, it is shown in [10] that the results for copper fluidisation are significantly closer to the model predictions when direct experimental values for the parameters u_t and n are employed in relation (4.2) rather than values from the standard correlations that are used to construct the global boundary of completely stable fluidisation shown in Fig. 7.1; the transition point itself is quite sharp and unambiguous in this case. For fluidisation of glass there is broad agreement with predictions: spheres of 0.5mm diameter and less exhibit essentially homogeneous fluidisation as predicted and, likewise, spheres with diameters in excess of 2mm produce complete voids. The transition point, however, is by no means sharp in this case so that reported values for ϵ_{mb} could be expected to show a considerable degree of scatter.

Intermediate diameter glass spheres fluidise in a singular manner that has been the subject of study and speculation: horizontal, high voidage bands are seen to form at, or close to, the distributor and to propagate through the bed; these were first reported by Hassett [14] and later by Anderson and Jackson [1] and El-Kaissy and Homsy [6]; these latter studies attributed the bands to voidage instabilities that reach a saturation level determined by non-linearities in the governing equations. An alternative explanation is considered in the following section.

8. Indeterminate Stability. For fluidised systems predicted to show a transition from stable to bubbling behaviour at a critical value of fluid velocity (and void fraction, ϵ_{mb}) the relative sharpness of the transition point can be examined with reference to the linearised equation (4.1) [11]. This is satisfied by the perturbation wave travelling at velocity ν:

$$\epsilon = A\exp(\alpha t + 2\pi i[z - \nu t]/\lambda),$$

$$\lambda = \frac{4\pi\nu}{B}\left(\frac{\nu^2 - u_\epsilon^2}{u_\epsilon^2 - \nu^2}\right)^{0.5}, \tag{8.1}$$

$$\alpha = \frac{B}{2\nu}(u_\epsilon - \nu).$$

The growth rate, α, for a key wave length, $\lambda = 20d_p$, is shown in Fig. 8.1 for three fluidised beds that are all predicted to transfer to the bubbling state at ϵ_{mb} equal to approximately 0.5. The relative magnitude of the growth rates is not sensitive to the particular wavelength chosen for illustrative purposes.

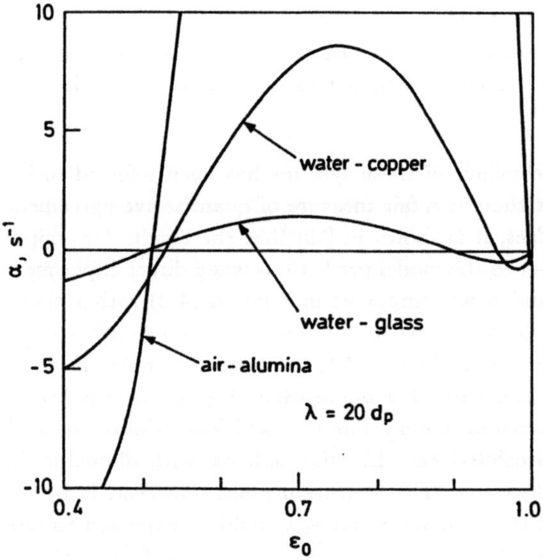

Fig. 8.1: Voidage perturbation growth rates as a function of void fraction for fluidised beds having ϵ_{mb} equal to approximately 0.5.

Fig. 8.1 shows that for the air-alumina system the perturbation growth rate changes near the transition point from a large negative to a large positive value over a very small expansion range: this is clearly indicative of a well defined transition from unambiguously stable to bubbling behaviour as observed in practice; a similar, if less extreme, situation occurs with water fluidisation of copper. For water fluidisation of less dense materials, such as glass, growth rates, whether negative or positive, remain relatively small suggesting that voidage perturbations created at the necessarily imperfect distributor, for example, could be expected to persist for extended periods giving rise to the band line formations observed in these systems. Further support for this view is provided by the measurements of perturbation wave frequency and velocity reported in [6] which are compared with the values predicted by eqn. (8.1) in Fig. 8.2: predicted values correspond to the continuity wave velocities for all observed frequencies.

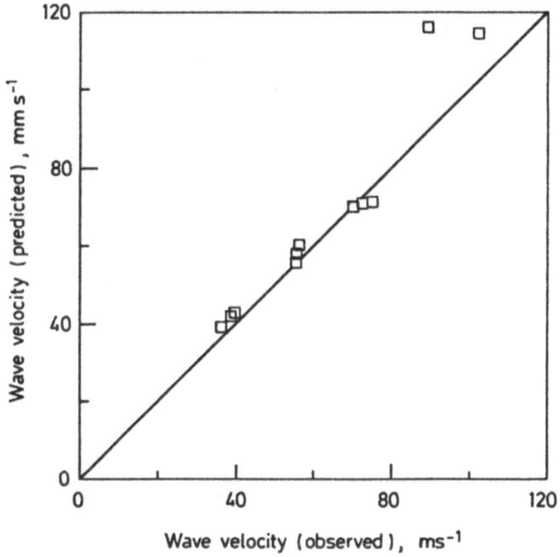

Fig. 8.2: Comparison of wave propagation velocities reported by El Kaissy and Homsy [6] with model predictions.

Finally, an augmented global map for fluidisation by ambient water is presented in Fig. 8.3: the shaded region indicates where, on the basis of relatively small key wavelength growth rates over a significant expansion region, persistent voidage inhomogeneities may be expected to occur [11].

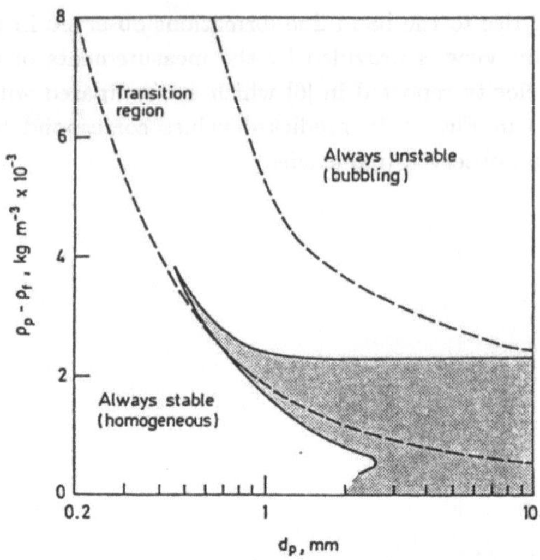

Fig. 8.3: Indeterminate stability in water fluidised beds: shaded region shows where sustained (i.e., slowly growing or decaying) voidage inhomogeneities may be expected.

Any predictions for low density particles must be regarded with extreme caution because both added mass effects and accelerational contributions to the fluid pressure field, which must be of significance in these cases, are not included in the model formulation. Nevertheless the indications of sustained voidage inhomogeneities for water fluidised, low density solids (such as plastics) are well confirmed by experiment.

REFERENCES

[1] T.B. Anderson and R. Jackson, *A fluid mechanical description of fluidised beds - comparison of theory and experiment*, Ind. Engng. Fundam., 8 (1969), pp. 137.

[2] J.M. Coulson and J.F. Richardson, *Chemical Engineering*, 3rd edn, 2, Pergamon Press, Oxford (1978).

[3] M.E. Crowther and J.C. Whitehead, *Fluidisation of fine particles at elevated pressure*, in Fluidisation, Cambridge University Press, Cambridge (1978), pp. 65–70.

[4] J.M. Dallavalle, *Micromeritics*, 2nd edn, Pitmans, London (1948).

[5] J.A.H. De Jong and J.F. Nomden, *Homogeneous gas-solid fluidisation*, Powder Technol., 9 (1974), pp. 91–97.

[6] M.M. El Kaissy and G.M. Homsy, *Instability waves and the origin of bubbles in fluidised beds*, Int. J. Multiphase Flow, 2 (1976), pp. 379–395.

[7] P.U. Foscolo and L.G. Gibilaro, *A fully predictive criterion for the transition between particulate and aggregate fluidisation*, Chem. Engng. Sci., 39 (1984), pp. 1667–1675.

[8] P.U. Foscolo and L.G. Gibilaro, *Fluid dynamic stability of fluidised suspensions: the particle bed model*, Chem. Engng Sci., 42 (1987), pp. 1489–1500.

[9] D. Geldart, *Types of gas fluidisation*, Powder Technol., 7 (1973) pp. 285–292.

[10] L.G. Gibilaro, I. Hossain and P.U. Foscolo, *Aggregate behaviour of liquid fluidised beds*, Can. J. Chem. Engng., 64 (1986), pp. 931–938.

[11] L.G. Gibilaro, R. Di Felice, P.U. Foscolo and S.P. Waldram, *Fluidisation quality: a criterion for indeterminate stability*, Chem. Engng. J., 37 (1988), pp. 25–33.

[12] L.G. Gibilaro, R. Di Felice and P.U. Foscolo, *On the minimum bubbling voidage and the Geldart classification for gas-fluidised beds*, Powder Technol., 56 (1988), pp. 21–29.

[13] L.G. Gibilaro, R. Di Felice, I. Hossain and P.U. Foscolo, *The experimental determination of one-dimensional wave velocities in liquid fluidised beds*, Chem. Engng. Sci., 44 (1989) pp. 101–107.

[14] N.J. Hassett, *The mechanism of fluidisation*, Brit. Chem. Engng., 11 (1961), pp. 777–780.

[15] K.V. Jacob and A.W. Weimer, *High-pressure particulate expansion and minimum bubbling of fine carbon powders*, AIChE.J., 33 (1987), pp. 1698–1706.

[16] S.N.P. Mutsers and K. Rietema, *Gas-solid fluidisation in a centrifugal field. The effect of gravity upon bed expansion*, Powder Technol., 18 (1977), pp. 249–256.

[17] P.N. Rowe, *A rational explanation for the behaviour of Geldart type A and B powders when fluidised*, AIChE. Annual Meeting, Miami Beach, Nov. 2–7 (1985), paper no. 58f.

[18] P.L. Slis, TH.W. Willemse and H. Kramers, *The response of the level of a liquid fluidised bed to a sudden change in the fluidising velocity*, Appl. Sci. Res., A8 (1959), pp. 209–217.

[19] G.B. Wallis, *One-dimensional waves in two-component flow (with particular reference to the stability of fluidised beds)*, United Kingdom Atomic Energy Authority, Report AEEW-R162 (1962).

[20] G.B. Wallis, *One-dimensional two-phase flow*, McGraw-Hill, New York (1979).

[21] R.H. Wilhelm and M. Kwauk, *Fluidisation of solid particles*, Chem. Engng. Prog., 44 (1948), pp. 201–218.

TRANSPORT PROCESSES IN CONCENTRATED SUSPENSIONS: THE ROLE OF PARTICLE FLUCTUATIONS

JAMES T. JENKINS* AND DAVID F. McTIGUE†

Abstract. Transport of momentum in slow flows of concentrated suspensions may be strongly dependent upon the fluctuations of particles about their mean motion. The intensity of the velocity fluctuations is an internal field that is the analog of temperature in classical kinetic theories. This viscous temperature is governed by a balance law that includes flux, production, and dissipation terms. We provide heuristic arguments to motivate the forms of the viscosity, conductivity, dissipation, and pressure in a theory that includes the viscous temperature. The approach parallels previous developments for dry, granular materials. Phenomena observed in flows of concentrated suspensions, including apparent normal stresses and shear-induced diffusion, are contained within the structure of this theory.

Key words. concentrated suspension, diffusion, effective viscosity, normal stress, rheology

1. Introduction. In 1954, Bagnold [1] reported results from a series of unique experiments on the rheology of concentrated suspensions. The measurements were made with a cylindrical Couette device, in which the fluid-filled inner cylinder had a flexible wall. Thus, Bagnold was able to measure not only the torque required to shear the suspension at a given rate, but also the normal thrust on the inner cylinder. The observations that have received the greatest attention were obtained at relatively high shear rates. Under these conditions, Bagnold measured both a shear traction, S, and a normal traction, N, on the inner cylinder that proved to be *quadratic* in the mean shear rate, v':

$$S = f_i \rho \sigma^2 (v')^2 \quad , \tag{1}$$

and

$$N = \beta_i f_i \rho \sigma^2 (v')^2 \quad , \tag{2}$$

where f_i is a strongly increasing function of ν, the volume fraction of solids, ρ is the mean mass density, σ is the particle diameter, and $\beta_i = N/S$ is a constant, observed by Bagnold to be about 3.3. He interpreted these results to be consequences of momentum transport via inertially-dominated, collisional interactions of particles. In recent years, Bagnold's experimental results in this regime have been corroborated by further measurements in annular shear cells [2, 3]. These observations have prompted a substantial effort directed toward understanding transport processes in particulate systems using elements of classical kinetic theory [*e.g.*, 4, 5].

*Department of Theoretical and Applied Mechanics, Cornell University, Ithaca, NY 14853. Partially supported by the U. S. Army Research Office through the Mathematical Sciences Institute at Cornell University.

†Fluid Mechanics and Heat Transfer Division I, Sandia National Laboratories, Albuquerque, NM 87185. Supported by the U. S. Department of Energy under contract DE-AC04-76DP00789 to Sandia National Laboratories.

At relatively low shear rates, a regime that has received somewhat less scrutiny, Bagnold observed shear and normal tractions *linear* in the shear rate:

(3) $$S = \mu f_v v' \quad ,$$

and

(4) $$N = \beta_v \mu f_v v' \quad ,$$

where μ is the interstitial fluid viscosity, f_v is again a strongly increasing function of the volume fraction of solids, and $\beta_v = N/S$ was supposed by Bagnold to be a constant, observed in his experiment to be about 1.3. The shear stress measurements (3) are generally consistent with the widely-applied concept of an effective viscosity, and indeed the function $f_v = S/\mu v'$ is conventionally referred to as the relative viscosity. It is the observation of a normal traction on the inner cylinder (4), apparently also linear in the shear rate, that is a rather unexpected and intriguing result. It certainly goes against experience with more familiar nonlinear fluids, where the first normal stresses to be observed are typically quadratic in the shear rate.[1] Furthermore, it can be argued [*e.g.*, 7] that, for creeping motions, all processes are exactly reversible, so that the pair distribution function for interacting particles in rectilinear shearing flows must be symmetric about the normal to the flow direction. Consequently, there should be no rate dependence in the pressure and *zero* normal stress differences.

Bagnold [1] presented extensive data for the inertial regime, but showed only one case that falls clearly in the linear, viscous regime. Nonetheless, he proposed an empirical form for the relative viscosity, $f_v(\nu)$, which he suggested is valid for a wide range of particle volume fractions, and he concluded that β_v is a constant. The paucity of data in the linear regime suggests caution in accepting Bagnold's results.

Recently, data have been obtained by Gadala-Maria [8] in a parallel-plate device in somewhat similar circumstances. Gadala-Maria was able to measure the normal thrust on the top plate, and also observed a normal traction linear in the shear rate at solid volume fractions of 0.3, 0.4, and 0.5. The coefficient corresponding to $\beta_v f_v(\nu)$ was observed to increase strongly with solid volume fraction, as did the effective viscosity, $f_v(\nu)$. However, in contrast to Bagnold's results, the normal traction was about one order of magnitude *less* than the shear traction. Furthermore, the ratio N/S increased with increasing volume fraction, taking the values 0.02, 0.15, and 0.16, respectively, for the three concentrations tested. Thus, Gadala-Maria's data seem to suggest that β_v must also be a function of ν.

It should be noted that the normal traction measured in the concentric-cylinder device and that recorded with the parallel-plate device have different relationships to the viscometric functions used to describe nonlinear rheology [*e.g.*, 9]. Bagnold's

[1] In *dilute* suspensions, normal stresses associated with small inertial effects have been predicted analytically [6], and are indeed quadratic in the shear rate.

device measures the total cross-stream normal stress, while Gadala-Maria's apparatus should record only effects of the normal-stress *differences*, and thus would not be sensitive to the isotropic part of the stress.

As is the convention in treating other nonlinear fluids, it is natural to seek a constitutive equation of the form $t_{ij} = -p\delta_{ij} + t^*_{ij}(\nu, d_{ij})$ that is consistent with (1) and (2) or (3) and (4); here, t_{ij} is the stress, p the pressure, t^*_{ij} the extra stress, $d_{ij} = (v_{i,j} + v_{j,i})/2$ the deformation rate, and v_i the velocity. Such attempts have been made, but one is quickly confronted with a physical contradiction. Bagnold's observation in both regimes was that the shear and normal stresses are *proportional.* This does not pose any difficulties in the simple shear flow his device approximated. However, consider a steady, plane, gravity flow between parallel walls in a vertical channel. In this situation, the shear stress varies linearly across the channel, and the normal stress is constant, *independent of constitutive behavior.* This is clearly in conflict with the requirement that the ratio of the shear and normal stresses is a constant.

This apparent paradox has been recognized for some time for the inertial regime [*e.g.*, 10]. A resolution has been found through consideration of the role of particle fluctuations in effecting momentum and energy transport. In the following sections, we review this argument briefly, in order to set the stage for an analogous argument for the viscous regime.

2. General Considerations. We assume that a granular material or the particle phase of a concentrated suspension[2] can be represented as a continuum fluid, for which the usual statements of balance of mass and momentum hold:

$$\dot{\rho} = -\rho v_{k,k} \quad , \tag{5}$$

and

$$\rho \dot{v}_i = t_{ik,k} + \rho b_i \quad , \tag{6}$$

where the overdot indicates a time derivative calculated following the mean motion, ρ is the mean density, v_k is the mean velocity, t_{ik} is the symmetric stress, and b_i is a force per unit mass.

A simple constitutive assumption for the stress is that for an isotropic fluid, in which we retain only terms linear in the rate of deformation,

$$t_{ik} = \left(-p + \overline{\lambda} d_{\ell\ell}\right)\delta_{ik} + 2\overline{\mu} d_{ik} \quad , \tag{7}$$

and anticipate that, in general, the pressure, p, and viscosity coefficients, $\overline{\lambda}$ and $\overline{\mu}$, are functions of the particle volume fraction, ν, as well as some measure of the particle velocity fluctuations. A convenient choice of the latter, denoted by T and called the granular temperature, is one-third the mean of their squared magnitude.

[2] We distinguish that latter from the former only by the importance of the interstitial fluid in transport processes.

It is evident from (5)–(7) that, in general, one must solve a given boundary-value problem for not only the density and velocity fields, but also the temperature field. Thus, an equation of balance for the mean fluctuation energy is also needed:

$$\frac{3}{2}\rho\dot{T} = -Q_{k,k} + t_{ik}v_{i,k} - \gamma \quad , \tag{8}$$

where Q_k is the flux of fluctuation energy, and γ is the collisional rate of dissipation of fluctuation energy per unit volume. Again, we anticipate that the flux can be written in the familiar form

$$Q_k = -\overline{\kappa}T_{,k} \quad , \tag{9}$$

where $\overline{\kappa}$ is a conductivity, which is again expected to depend, in general, upon ν and T.

Consider now steady, rectilinear, shearing flows, $v = v_1(x_2)$, in which the mean velocity, volume fraction, and temperature are supposed to vary only in the cross-stream direction. In this case, the mass balance is satisfied identically, and the streamwise and cross-stream components of the momentum balance and the energy balance reduce, respectively, to:

$$0 = (\overline{\mu}v')' + \rho b_1 \quad , \tag{10}$$

$$0 = -p' + \rho b_2 \quad , \tag{11}$$

and

$$0 = (\overline{\kappa}T')' + \overline{\mu}(v')^2 - \gamma \quad , \tag{12}$$

where a prime indicates a cross-stream derivative.

In order to complete the model for steady, shearing flows, it is now necessary to specify functional forms for the pressure, p, viscosity, $\overline{\mu}$, conductivity, $\overline{\kappa}$, and dissipation, γ. Some simple motivational arguments for these quantities in both the inertial and viscous regimes are presented in the following two sections.

3. The Inertial Regime. In the inertial regime, we assume the transport processes are dominated by collisional interactions among the particles and focus attention on the role of the velocity fluctuations. We summarize very briefly here the heuristic arguments offered by Haff [4] to motivate functional forms for the constitutive relations of interest (Table 1).

Consider a rather concentrated system of identical, smooth, nearly elastic spheres, and neglect the presence of an interstitial fluid. The mean mass density is related to the particle properties by $\rho = nm$, where n is the number density of particles, and m is the mass per particle. For a dense system, the particles remain in close proximity to one another, and the mean separation distance, h, is much less than the particle diameter, σ, *i.e.*, $h/\sigma \ll 1$. The density then scales like $\rho \sim m/\sigma^3$.

A particle fluctuates with a velocity that has a magnitude of order $T^{1/2}$. In a typical collision, then, the particle will exchange momentum of the order of $mT^{1/2}$. The rate of collisions goes like $T^{1/2}/h$, and the area over which the resulting flux occurs is proportional to $1/\sigma^2$. Combination of these relations yields the expected scaling for the pressure, or momentum flux due to the fluctuating motion,

$$p \sim \rho \frac{\sigma}{h} T \quad . \tag{13}$$

The viscosity is a measure of the mean cross-stream flux of streamwise momentum. The mean streamwise momentum of a particle that is exchanged in a collision with a neighboring particle is $m\Delta v$, where $\Delta v \sim \sigma v'$ is the relative mean velocity of particles in a velocity gradient v'. The collision rate and the area scale as before. Consequently, the momentum flux, or shear traction, S, is, up to a factor,

$$S \sim \rho \frac{\sigma}{h} \sigma T^{1/2} v' \quad . \tag{14}$$

Thus,

$$\overline{\mu} \sim \rho \frac{\sigma}{h} \sigma T^{1/2} \quad . \tag{15}$$

A similar argument can be made for the form of $\overline{\lambda}$, the bulk viscosity in (7).

The fluctuation kinetic energy transfer per collision is proportional to $m\Delta T$, where $\Delta T \sim \sigma T'$ is the difference in the mean squared fluctuation velocity between neighboring particles. Multiplied by the collision frequency, $T^{1/2}/h$, and divided by the area scale, σ^2, this gives the transverse flux of fluctuation energy, Q,

$$Q \sim \rho \frac{\sigma}{h} \sigma T^{1/2} T' \quad , \tag{16}$$

and the conductivity,

$$\overline{\kappa} \sim \rho \frac{\sigma}{h} \sigma T^{1/2} \quad . \tag{17}$$

Finally, because the particles are inelastic, kinetic energy is lost in each collision. The energy loss is characterized by a coefficient of restitution e that varies between zero and one. The product of the loss per collision, $\sim (1-e)mT$, the collision frequency, $\sim T^{1/2}/h$, and the number density, $\sim 1/\sigma^3$, yields the form of the volumetric dissipation rate, γ,

$$\gamma \sim \rho \frac{\sigma}{h} \frac{(1-e)}{\sigma} T^{3/2} \quad . \tag{18}$$

Consider now the special case of a *homogeneous* shear flow, in which v', ν (which, presumably, can be related to the mean separation h by a geometrical or statistical argument), and T are constant. The flux of mean fluctuation energy, Q (16), must then vanish, and the energy equation (12) reduces to a balance between production by the mean shearing motion, $\overline{\mu}(v')^2$, and dissipation through inelastic losses, γ. This gives an algebraic relationship for the temperature in terms of the mean motion,

$$T \sim \frac{\sigma^2}{1-e}(v')^2 \quad . \tag{19}$$

Substitution of (19) into (14) shows that, in this special case, the shear stress is quadratic in the shear rate, v'. In addition, substitution of (19) into (13) shows that the normal stress, $N = p$, is also quadratic in v'. Both results are in accord with equations (1) and (2), based on the experimental observations of Bagnold [1] and others. The ratio $\beta_i = N/S$ is seen to scale with $(1-e)^{-1/2}$. The more detailed, kinetic theory of Jenkins and Savage [5] yields explicit numerical values for the coefficients missing from (13), (15), (17), and (18). They find, for example, $\beta_i = N/S = \sqrt{5\pi/12}(1-e)^{-1/2}$. Bagnold observed that $\beta_i \simeq 3.3$, indicating, in the context of this model, that the coefficient of restitution characterizing his material was $e \simeq 0.9$.

It is important to note that these simple relationships between stress and mean shear rate do not hold in general. In a general, steady, rectilinear, shear flow, v', ν and T are not constant across the flow, and the energy and momentum balances must be solved simultaneously to determine the stress fields. Indeed, this is fortunate, for it removes the contradiction noted above for flow between plane, parallel walls. In that case, because of the term involving the energy flux in (12), T need not vanish at the centerline, and $N = p$ as given by (13) can remain finite there, even though symmetry requires that the shear traction, S, must vanish.

4. The Viscous Regime. We propose here that, in concentrated suspensions, particle fluctuations play a role entirely analogous to that in the inertial regime, and can dominate the transport processes at sufficiently high concentration. In this case, there is still interaction between particles, but it is effected through the interstitial viscous fluid, rather than through collisions. The forces of interaction are then due to the local pressure and shear stresses at the fluid/particle interfaces. At high concentrations, these forces are dominated by the lubrication forces in the thin, interparticle, fluid films. The same equations of balance for the mean mass (5), momentum (6), and fluctuation energy (8) fields hold, and constitutive equations for the momentum (7) and energy (9) fluxes of the same form are assumed.

A simple dimensional argument points toward the necessary forms for the pressure, viscosity, conductivity, and dissipation. At low particle Reynolds numbers, $R_p = \rho T^{1/2}\sigma/\mu$ (where μ is the interstitial fluid viscosity), inertial exchange of momentum in collisions gives way to forces of viscous interaction. We then expect that the density, ρ, cannot enter into the theory, but must be replaced by $\mu/(\sigma T^{1/2})$; *e.g.*, the particle-scale momentum flux scales like ρT in the inertial regime, but like $\mu T^{1/2}/\sigma$ in the viscous regime. This leads to scalings as shown in Table 1. In the following paragraphs, we provide heuristic arguments in the same spirit of those given for the inertial regime.

The magnitude of the force of interaction, F, due to two identical spheres in close approach along their line of centers at a relative velocity U is, at leading order in the lubrication analysis [*e.g.*, 11],

$$F \sim \mu\sigma U \frac{\sigma}{h} \quad . \tag{20}$$

The relative velocity of interacting particles is due to both the local velocity fluctuations and the mean shear. Thus, there are two parts to U, one scaling like $T^{1/2}$, and the other scaling like $\Delta v \sim \sigma v'$. Again, the number of particles per unit area is proportional to $1/\sigma^2$, so we obtain from the mean shear a cross-stream flux of streamwise momentum,

$$S \sim \mu \frac{\sigma}{h} v' \quad , \tag{21}$$

and an effective viscosity,

$$\overline{\mu} \sim \mu \frac{\sigma}{h} \quad . \tag{22}$$

As before, a similar argument can be made for the form of the bulk viscosity, $\overline{\lambda}$. Note that the effective viscosity (22) does not depend on the fluctuation energy, T; it is a function of the volume fraction only, which is consistent with a large body of previous work [*e.g.*, 12].

The flux of mean fluctuation energy, Q, is the difference in the rate of working of the interparticle force (20), $F \sim \mu\sigma T^{1/2}$, associated with the increment in the fluctuation velocity, $\Delta T^{1/2} \sim (\sigma/T^{1/2})T'$. Multiplied by the number of particles per unit area, $\sim 1/\sigma^2$, this yields a cross-stream flux

$$Q \sim \mu \frac{\sigma}{h} T' \quad , \tag{23}$$

and an effective conductivity

$$\overline{\kappa} \sim \mu \frac{\sigma}{h} \quad . \tag{24}$$

The viscous dissipation rate in a single, interparticle gap is the rate of working of the fluctuation in force, $\sim \mu\sigma^2 T^{1/2}/h$, through the fluctuation in relative velocity, $T^{1/2}$. The number density of particles scales with $1/\sigma^3$. Therefore, the average rate of dissipation of fluctuation energy per unit volume is expected to be

$$\gamma \sim \mu \frac{\sigma}{h} \frac{1}{\sigma^2} T \quad . \tag{25}$$

The pressure is the mean of the fluctuations in force over a unit area. Consequently, a non-vanishing pressure can be obtained only if there is some asymmetry in the force; for example, if the force of departures is a fraction e of the force of approach. Leighton and Acrivos [13] note that particle contact at asperities is likely to provide such an asymmetry. When it is present, the pressure scales like

$$p \sim \mu \frac{\sigma}{h} \frac{(1-e)}{\sigma} T^{1/2} \quad . \tag{26}$$

Calculations analogous to those of Jenkins and Savage [5] in the inertial regime confirm the scalings and provide the coefficients [14].

Consider again a steady, homogeneous, shear flow; the balance of mean fluctuation energy (12) again reduces to an algebraic relation between the shear rate, v', and the temperature, T. This indicates that

$$T \sim \sigma^2 (v')^2 \quad , \tag{27}$$

so that the normal traction, $N = p$ given by (26), appears to be linear in the shear rate, just as Bagnold observed (*cf* eq. 4). Thus, with $N \sim \mu\sigma(1-e)v'/h$ and $S \sim \mu\sigma v'/h$, we find that $\beta_v = N/S$ is a constant. We again emphasize that this is a special case; in general, the temperature field, and thus the pressure, is coupled in a much more complicated way to the mean shearing through the energy equation.

Table 1. Inertial *vs* viscous transport.

	Inertial	Viscous
Pressure, p	$\sim \rho\frac{\sigma}{h}T$	$\sim \mu\frac{\sigma}{h}\frac{(1-e)}{\sigma}T^{1/2}$
Viscosity, $\overline{\mu}$	$\sim \rho\frac{\sigma}{h}\sigma T^{1/2}$	$\sim \mu\frac{\sigma}{h}$
Conductivity, $\overline{\kappa}$	$\sim \rho\frac{\sigma}{h}\sigma T^{1/2}$	$\sim \mu\frac{\sigma}{h}$
Dissipation, γ	$\sim \rho\frac{\sigma}{h}\frac{(1-e)}{\sigma}T^{3/2}$	$\sim \mu\frac{\sigma}{h}\frac{1}{\sigma^2}T$

5. Shear-induced Self-Diffusion. Leighton and Acrivos [13, 15] describe several phenomena observed in the course of viscometric measurements on concentrated suspensions that they ascribe to self-diffusion of particles in shearing flows. In various configurations, they observe evidence for a slow, net drift of particles from regions of high concentration to regions of low concentration.

Consider a plane, rectilinear, shear flow as defined in equations (10)–(12), oriented so that gravity is transverse to the flow. The magnitude of the gravitational force per unit volume on the particles is $\nu(\rho_s - \rho_f)g$, where ρ_s and ρ_f are the mass densities of the solid and fluid phases, respectively. If the particles have a small mean velocity w transverse to the flow, there is an additional volume force due to viscous drag. The magnitude of this force is given in terms of a hindrance function

$H(\nu)$ as $18\mu\nu w/\sigma^2 H$ [*e.g.*, 16]. In this case, (11) can be written as

$$p' = -\nu g \Delta\rho - \frac{18\mu\nu}{\sigma^2 H} w \quad . \tag{28}$$

In the absence of gravity, the diffusive flux is related to the pressure gradient by

$$\nu w = -\frac{\sigma^2 H}{18\mu} p' \quad . \tag{29}$$

In a flow close to uniform shear, we ignore gradients in the temperature and define a diffusivity $\mathcal{D}$ by

$$\mathcal{D} = \frac{\sigma^2 H}{18\mu} \frac{\partial p}{\partial \nu} \quad , \tag{30}$$

so that $\nu w \simeq -\mathcal{D}\nu'$. Equation (26) indicates the ν derivative of p is proportional to $\mu(1-e)T^{1/2}/\sigma$, and, from (27), $T^{1/2} \sim \sigma v'$. Thus,

$$\mathcal{D} \sim \sigma^2 v'(1-e) f \quad , \tag{31}$$

where f is a function of ν alone. This is the scaling observed by Leighton and Acrivos [13, 15].

With gravity present and the flux zero [17], the steady concentration distribution is determined by solving

$$p' = -\nu g \Delta\rho \tag{32}$$

in conjunction with (10) and (12).

6. Summary and Discussion. We suggest that certain unusual phenomenology exhibited by concentrated suspensions can be captured by a model that takes account of the fluctuations of particles about their mean motion. The model follows closely the structure of recent theories for dry, granular materials. In these theories, particle fluctuations are characterized by a scalar field analogous to temperature in classical kinetic theory. An equation of state relates the pressure to the density (*i.e.*, volume fraction) and temperature (*i.e.*, fluctuations), and functional forms for the effective viscosity, conductivity, and dissipation rate are also required. For a dry granular material, the forms of these functions can be motivated by consideration of the fluxes of momentum and fluctuation kinetic energy through collisional particle interactions. In the presence of a viscous, interstitial fluid, at small particle Reynolds numbers, these exchanges must be effected via the fluid. Simple scaling arguments are proposed to motivate the forms of the required constitutive relations. Each of the quantities p, $\overline{\mu}$, $\overline{\kappa}$, and γ, must be of lower order in the temperature by one half power.

In a homogeneous, shear flow, a simple, dimensional argument shows that the temperature must scale with $(\sigma v')^2$. Thus, in the inertial regime, where $\overline{\mu} \sim T^{1/2}$, the shear stress is quadratic in the shear rate, and the pressure, which is linear in T, likewise is quadratic in the shear rate, as observed in experiments. In the viscous regime, the viscosity does not depend upon T, and the pressure, which is proportional to $T^{1/2}$, is linear in the shear rate, again in accord with experimental observations.

The introduction of the viscous temperature and its associated balance law seems to provide a structure that permits interpretation of the phenomena observed by Leighton and Acrivos [13, 15, 17], at least when the particle interactions are asymmetric and a particle pressure exists. Direct measurements of the pressure in shearing flows of a concentrated suspension are necessary in order to determine whether the predicted dependence on the shear rate through the viscous temperature is present.

REFERENCES

[1] R. A. Bagnold, *Experiments on a gravity-free dispersion of large solid spheres in a Newtonian fluid under shear*, Proc. Royal Soc. London, A223 (1954), pp. 49–63.

[2] S. B. Savage and M. Sayed, *Stresses developed by dry cohesionless granular materials sheared in an annular shear cell*, J. Fluid Mech., 142 (1984), pp. 391–430.

[3] D. M. Hanes and D. L. Inman, *Observations of rapidly flowing granular-fluid materials*, J. Fluid Mech., 150 (1985), pp. 357–380.

[4] P. K. Haff, *Grain flow as a fluid mechanical phenomenon*, J. Fluid Mech., 134 (1983), pp. 401–433.

[5] J. T. Jenkins and S. B. Savage, *A theory for the rapid flow of identical, smooth, nearly elastic, spherical particles*, J. Fluid Mech., 130 (1983), pp. 187–202.

[6] C.-J. Lin, J. H. Peery, and W. R. Schowalter, *Simple shear flow round a rigid sphere: inertial effects and suspension rheology*, J. Fluid Mech., 44 (1970), pp. 1–17.

[7] J. F. Brady and G. Bossis, *The rheology of concentrated suspensions of spheres in simple shear flow by numerical simulation*, J. Fluid Mech., 155 (1985), pp. 105–129.

[8] F. A. Gadala-Maria, *The rheology of concentrated suspensions*, Ph.D. thesis, Stanford University, 1979.

[9] W. R. Schowalter, *Mechanics of Non-Newtonian Fluids* Pergamon, Oxford, 1978.

[10] J. T. Jenkins and S. C. Cowin, *Theories for flowing granular materials*, Mechanics Applied to the Transport of Bulk Materials, S. C. Cowin, ed., Am. Soc. Mech. Eng., AMD–31 (1979), pp. 79–89.

[11] D. J. Jeffrey and Y. Onishi, *Calculation of the resistance and mobility functions for two unequal rigid spheres in low-Reynolds-number flow*, J. Fluid Mech., 139 (1984), pp. 261–290.

[12] D. J. Jeffery and A. Acrivos, *The rheological properties of suspensions of rigid particles*, AIChE J., 22 (1976), pp. 417–432.

[13] D. Leighton and A. Acrivos, *Measurement of shear-induced self-diffusion in concentrated suspensions of spheres*, J. Fluid Mech., 177 (1987), pp. 109–131.

[14] J. T. Jenkins and D. F. McTigue, *Viscous fluctuations and the rheology of concentrated supensions*, in preparation.

[15] D. Leighton and A. Acrivos, *The shear-induced migration of particles in concentrated suspensions*, J. Fluid Mech., 181 (1987), pp. 415–439.

[16] R. H. Davis and A. Acrivos, *Sedimentation of noncolloidal particles at low Reynolds numbers*, Ann. Rev. Fluid Mech., 17 (1985), pp. 91–118.

[17] D. Leighton and A. Acrivos, *Viscous resuspension*, Chem. Eng. Sci., 41 (1986), pp. 1377–1384.

STRESS IN DILUTE SUSPENSIONS

STEPHEN L. PASSMAN*

Abstract. Generally, two types of theory are used to describe the field equations for suspensions. The so-called "postulated" equations are based on the kinetic theory of mixtures, which logically ought to give reasonable equations for solutions. The basis for the use of such theory for suspensions is tenuous, though it at least gives a logical path for mathematical arguments. It has the disadvantage that it leads to a system of equations which is underdetermined, in a sense that can be made precise. On the other hand, the so-called "averaging" theory starts with a determined system, but the very process of averaging renders the resulting system underdetermined. I suggest yet a third type of theory. Here, the kinetic theory of gases is used to motivate continuum equations for the suspended particles. This entails an interpretation of the stress in the particles that is different from the usual one. Classical theory is used to describe the motion of the suspending medium. The result is a determined system for a dilute suspension. Extension of the theory to more concentrated systems is discussed.

1. Introduction. In theories of multiphase flows, it is natural to postulate or to derive equations of balance similar to those occurring in the theory of dilute mixtures of gases [1,2]. The usual process of doing so, along with reasonable assumptions for the constitutive properties of the materials composing the flow, always leads to a system with more unknowns than equations. Though there is no definitive reason that this is a bad situation, intuition abetted with proved theorems for special types of systems indicates that the normal desirable situation is the same number of equations as unknowns. The resulting quandary for multiphase flows is known as the "closure problem", and methods for "solving" or "closing" it, that is, finding "sufficient" additional equations, has been the focus of considerable research in multiphase flows. Here, we try to shed some light on such problems. Essential to doing so is stating the problems unequivocally. In order to do that, we choose a special but interesting physical situation, then give typical equations of balance and constitutive equations for that physical situation, according to a continuum theory and an averaging theory. The closure problem occurs in both types of theories, though its form is different. However, it is possible to formulate theories in which the closure problem does not occur, and therefore need not be solved. A physical basis for such a system is presented, and a putative set of field equations is suggested.

2. Determined, Underdetermined, and Overdetermined Systems of Equations. Assume we have a system of equations of the form

$$f_i(y_j, D_k y_j) = 0,$$

with $i = 1, \ldots, n$; $j = 1, \ldots, m$; and $k = 1, \ldots, p$. The f_i are n functions of the m variables y_j and their derivatives up to order p. I will call the system *determined* if $n = m$, *overdetermined* if $n > m$, and *underdetermined* if $n < m$. By our definition,

*Pittsburgh Energy Technology Center, Pittsburgh PA 15236, on temporary assignment from Sandia National Laboratories, Albuquerque NM 87185.

all systems of equations are of one of these three types. Ideally, of course, it would be convenient if determined systems always had solutions and they were unique, if overdetermined systems never had solutions, and if underdetermined systems always had families of solutions. That this is not the case can be shown by examples. To begin, consider the underdetermined system

$$x_1^2 + x_2^2 = 0. \tag{1}$$

Naturally, specifying a system of equations is meaningless without specifying their domain, but since this paper is informal, I follow the convention of doing so tacitly, that is, all functions are mappings of all real variables for which they can be defined reasonably into a range defined by the function. Here, the underdetermined algebraic system (1) has the single unique solution

$$x_1 = 0, \quad x_2 = 0.$$

Now consider the determined system

$$\begin{aligned} x_1 + x_2 &= 1, \\ 2x_1 + 2x_2 &= 2. \end{aligned} \tag{2}$$

This system does not have a unique solution, rather it has an infinite one-parameter family of solutions. Finally, consider the overdetermined system

$$\begin{aligned} x_1 + x_2 &= 1, \\ 2x_1 + 2x_2 &= 2, \\ 2x_1 + 4x_2 &= 4. \end{aligned} \tag{3}$$

This system has a unique solution. All of the examples cited involve algebraic equations, not differential equations, but of course examples of the same type can be constructed with differential systems.

It is easy to object to the arguments above because the systems cited are "special" or "pathological". Indeed, I agree with that type of objection, and in a sense that is just the point of this discussion, for to make such arguments, one must use very special properties of the systems. Furthermore, for each system, it would have been possible to rearrange the system using simple manipulations, obtaining very complicated new systems with *exactly the same properties*. Proofs of existence or non-existence, uniqueness or non-uniqueness, then would be much more elaborate exercises, perhaps depending on sophisticated mathematics or luck for ultimate outcome.

A different line of argument is possible. Most often, the equations governing multiphase flow are systems of partial differential equations, so complicated that they are not easily amenable to existence or uniqueness theorems. Such equations may be relevant to important problems in technology, so important that they must be solved, in no matter in how vague a sense, immediately. More often than not, that means the use of large computer codes. Despite the above examples, we find that an intrinsic part of building such codes is the desire of the numerical analyst for determined systems, that is, ***systems determined in exactly the sense defined here***. The idea that the equations obtained by specialists in multiphase flow are "independent" is supported in some vague sense by the fact that they have different physical meanings. For example, some are balance equations for the constituents, some are constitutive equations, and some are constraints.

3. Continuum Theories of Multiphase Flows. Here, we consider a standard type of theory for multiphase flows, as derived from continuum considerations.[1] Intrinsic to such considerations is the assertion that each constituent fills all of a region of space. This is the basic assumption of theories of interpenetrating continua or "solutions" [4]. The theory then is made to model a multiphase medium by the inclusion of volume fractions ϕ_a as basic variables. A typical set of field equations for such a continuum having n constituents is

$$
\begin{aligned}
\Sigma_{a=1}^{n}\phi_a &= 1, \\
\dot{\phi}_a + \phi_a \operatorname{div} \mathbf{v}_a &= 0, \\
\rho_a \dot{\mathbf{v}}_a = \rho_a \mathbf{b}_a + \mathbf{m}_a &+ \operatorname{div} \mathbf{T}_a, \\
\Sigma_{a=1}^{n}\mathbf{m}_a &= 0, \\
\{\mathbf{T}_a, \mathbf{m}_a\} = g(\mathbf{v}_b, \phi_b, &\text{ and their derivatives}).
\end{aligned} \tag{4}
$$

Here for simplicity we consider only pure mechanics, with m_a the forces of interaction between constituents, and the other symbols have the obvious meanings. The first equation expresses the fact that the material is saturated; the second and third are balances of mass and momentum for each of the constituents. Each constituent is assumed to be incompressible, and p_a are the reactions to those constraints. The fourth equation is conservation of momentum for the mixture, and the last equations express constitutive properties of the constituents, in particular the dependence of the stress on the deformation rate and other properties for each constituent, and appropriate expressions for interactions of the two materials. We note that these last expressions can be somewhat problematical, and in fact debate about their forms has generated a considerable literature. They are not discussed here. Rather, we assume they are known for applications of particular interest.[2] Our intuitive feeling from considering the physics of this situation is that such a system of equations is "complete", but in fact that is not the case. Here and henceforth, for the purpose of counting equations, we assume the multiphase flow consists of two constituents. Though such is not always the case, for the purpose of our arguments here, that case is general. The result is a system of 9 equations in the 10 unknowns $\{p_a, \phi_a, \mathbf{v}_a\}$, that is, the system is *underdetermined*. The usual physical motivation for this apparent quandary is plausible: Though Equations (4) express the exchange of momentum between constituents, that is not the only way

[1] Discussion of the basis of such theories, as well as references to the standard works, are given in [3].

[2] See [5] and [6] for a discussion of these equations.

the constituents interact, for ***in addition***, there should be a force balance between the constituents. The most primitive visualization of this is sufficient for arguments here. That is, the solid phase is considered to consist of spherical particles of one size, surrounded by the fluid phase. Then a radial force balance on a single particle gives

$$p_s = p_f. \tag{5}$$

This obvious and elegant closure argument gives a system of equations which is notoriously ill-behaved [7,8], so much so that it must be rejected. More sophisticated arguments are possible, and they sometimes appear to suffice to render the system of equations thus obtained to be at least well-behaved enough to be handled by standard computational techniques. Usually, the arguments adduced are generalizations of those leading to Equation (5) in that they consider a particle in a flow field of a known type at infinity, then use techniques of hydrodynamics to solve or partially solve for the flow field around the spherical particle. Surface tension may be considered also. Of course the resulting pressure on the particle is a function of position on its surface relative to the flow field at infinity, so some sort of integration is required. The result is

$$p_s = p_f + f(\Pi, \sigma), \tag{6}$$

where Π denotes properties of the flow field, and σ denotes properties sufficient to characterize surface tension. I note that since f depends upon the flow field at infinity, adducing it as a constitutive relation valid for all flows has the potential for leading to inaccurate results.[3]

In addition to the mathematical argument against using Equation (5) as a closure relation, there is a physical argument against it, which also is inherited by Equation (6). In doing the arguments leading to these equations, the assumption is that p_s and p_f are pressures "in" the respective materials, and that it is appropriate to write an expression for one in terms of the other. The need for the closure relation comes from arguments about the system (4). In this system, the pressures are derived as reactions to constraints. Therefore [10], they are dependent variables of the system of equations, totally independent of one another. In treating the complete system of equations and boundary conditions,[4] the quantities p_s and p_f thus cannot be related ***a priori***. Another way to see this is that in fact Equations (4) are field equations for the ***whole continuum***, while the closure relation (6) is not. Rather, it is derived from a "micromechanical" argument, then scaled up in a way the nature of which never is made clear, so that the symbols p_s and p_f in Equations (6) are ***assumed*** to have the same meaning as the same symbols in the closure relation, without proof or explanation.[5]

---

[3] A similar difficulty arises in rheology, where it is commonly known [9] that no constitutive equation giving an accurate representation of the physics of shearing flows also represents stretching flows adequately.

[4] Boundary conditions in themselves constitute a difficult problem for multiphase flows. They are not discussed in this paper.

[5] Of course, the same argument can be made for the expressions in (4) for $\mathbf{m}_a$. There, however,

4. Averaging Theory. The basic ideas behind averaging theories are diametrically opposite from that of the continuum theories, though the objective—finding differential equations for fields, valid throughout a body—is exactly the same. It is perhaps fortuitous, or perhaps a sign that the equations actually represent some sort of physical "truth", that the forms of the equations resulting from the two approaches are so similar. For averaging theories, the region of space occupied by the material is thought of as being occupied by two different types of body, the suspended particles and the suspending medium. Each of the types of body is considered to be distinct in the sense that the union [11] of the bodies constitutes all of the space occupied by the composite body, while the intersection is empty. Then each of the types of body is an ordinary continuum, and satisfies exactly the balance and constitutive laws expected of an ordinary continuum, that is,

$$\begin{aligned} \operatorname{div} \mathbf{v}_a &= 0, \\ \rho_a \dot{\mathbf{v}}_a &= \rho_a \mathbf{b}_a + \operatorname{div} \mathbf{T}_a, \\ \{\mathbf{T}_a\} &= g(\mathbf{v}_b, \rho_b \text{ and their derivatives}). \end{aligned} \tag{7}$$

Here, of course, the bodies still are capable of momentum interaction, but unlike the previous situation, the micromechanical model for momentum interaction has a clear meaning. This is eight equations in eight unknowns $(\mathbf{v}_a, p_a)$, and thus is a determined system. Moreover, conditions for the difference of pressure such as Equation (6) now have a correct theoretical status, for now *they are not field equations, rather, they are boundary conditions.* Thus a determined system is obtained, and it is mathematically correct and physically plausible. The difficulty, of course, is that to formulate a boundary-value problem, a reasonable set of boundary and initial values for every particle in the system is needed. Such information normally is not available for any physical problem. Even if it were, finding a solution, with or without a computer as an intermediary, would be a nearly hopeless task. Moreover, even if such a solution were found, most of the information it contained would be of little use, because it would be too detailed. A plausible way to digest such data would be to average it in some sense. The usual approach in averaging theory is, not to go to the considerable trouble of averaging the solutions to (7), but rather to average (7) and then solve the averaged equations. Such an approach is highly appealing intellectually, but is fraught with mathematical difficulty. This paper is not the place to discuss such difficulties in detail. One, of course, is that the term "average" which has been used in a very vague way here, must be given a precise meaning. It is fortuitous that, for most of the averaging methods tried so far, the resulting equations have almost the form of the Equations (4) derived from the continuum theory. Unfortunately, for every method of averaging I have seen, though a determined system is averaged, the result of the averaging process is an underdetermined system, that is, the averaging process makes the closure problem reappear. Most readers will be familiar with why this happens without going through the

---

the status of the equations is clear, because the appropriate micromechanical arguments can be used as *motivation* for the continuum theory, which then gives exact relations having the same status as field equations.

details, for the averaging process always is similar to that used in turbulence theory and some of the extra terms are of the same form as Reynolds stresses. In a broad sense, then, though the continuum theory and the averaging theory start from different places and proceed by different methods, they end in approximately the same place: underdetermined systems of approximately the same form.

5. Sketch of a Theory for Dilute Suspensions. Previously in this paper, much has been made of the fact that most of the equations in the continuum theory have been "postulated". It is possible to interpret that terminology as meaning that they have been made up with no mathematical or physical basis. In fact, that is far from true. The kinetic theory of dilute monatomic gases for identical gas molecules is well-known and is commonly taught in courses for graduate students in science and engineering [1,2,12]. Much less well known is the fact that there is a similar theory for gases where there are a finite number of different types of molecules; in other words, a solution of several gases [13,14]. The resulting balance equations are exactly identical to those for the postulated theory of mixtures.

Here, I use the motivation of the kinetic theory of gases to support a mixture theory in an entirely different way. Most important for the discussion here is the fact that there is an exact definition of the pressure, and it is not the pressure "in" the particles, rather it is a momentum flux — an entirely different concept.[6] Moreover, it is possible to force agreement of the theory with that of a viscous compressible gas, with the viscosity determined in terms of molecular parameters. I consider a dilute solution of particles in an inviscid fluid. Consider only the particles. They are an agglomeration of molecules, exactly like those in the theory of a monatomic gas, except that the scale of the molecules is somewhat larger than in a gas. Thus, precisely the same arguments can be used to motivate a continuum theory for the particle phase of the multiphase flow as is used for a gas. All of the expressions are the same, and *e.g.*, one can accept the viscosity of the particle phase as a phenomenological coefficient, or one can consider it to be determined from molecular quantities, according to one's taste. In either case, unlike in the theories discussed in the previous two sections of this paper, it does have meaning. The equations for the particle phase then are

$$
\begin{aligned}
\dot{\rho}_s + \rho_s \operatorname{div} \mathbf{v}_s &= 0, \\
\rho_s \dot{\mathbf{v}}_s &= \rho_s \mathbf{b}_s + \mathbf{m} + \operatorname{div} \mathbf{T}_s, \\
p_s &= \hat{p}_s(\rho_s), \\
\mathbf{T}_s &= -p_s(\rho_s)\mathbf{1} + \hat{\mathbf{T}}_s(\text{sym grad}\, \mathbf{v}_s).
\end{aligned}
\tag{8}
$$

This is a system of 5 equations in 5 unknowns, that is, a determined system. Now let the molecules be submerged in an incompressible fluid. Naturally, there will be an interaction between the particles and the fluid, and this interaction can be expressed as a constitutive equation for **m**, which can be thought of as a part of the

[6] In another paper in this volume, O. Walton uses the same definition in his computer molecular dynamics simulations.

body force $\mathbf{b}_f$. Of course, the equations for the fluid phase are the expected ones,

$$\begin{aligned} \operatorname{div} \mathbf{v}_f &= 0, \\ \rho_f \dot{\mathbf{v}}_f &= \rho_f \mathbf{b}_f - \mathbf{m} + \operatorname{div} \mathbf{T}_f, \\ \mathbf{T}_f &= \hat{\mathbf{T}}_f(\text{sym grad}\, \mathbf{v}_f). \end{aligned} \tag{9}$$

again, a determined system. Thus, for a theory of this type, no closure problem exists.[7]

Generally in a theory of this type, one expects to see volume fractions appear intrinsically. Since the ideas here are for a very dilute, saturated suspension, the concept is not very important, except, perhaps, in the constitutive equation for $\mathbf{m}$ and in formulas for the "effective viscosity" [5,12]. Of course the idea can be introduced formally by setting

$$\rho_f = \gamma_f \phi_f,$$

with γ_f the constant mass per unit volume of fluid, and

$$\phi_f + \phi_s = 1.$$

These substitutions introduce the same number of equations as unknowns.

Acknowledgments. Donald Drew, R.C. Givler, David McTigue, Mehrdad Massoudi, and Kathleen Pericak-Spector have been kind enough to discuss the ideas presented here with me. The work was supported by the United States Department of Energy. I express my deep appreciation to the Institute for Mathematics and its Applications for inviting me to lecture on this subject, and for help in preparation of the manuscript.

REFERENCES

[1] C. Truesdell and R.G. Muncaster, *Fundamentals of Maxwell's Kinetic Theory of a Simple Monotomic Gas*, Academic Press, New York, 1980.

[2] W.G. Vincenti and C.H. Kruger, Jr., *Introduction to Physical Gas Dynamics*, Wiley, New York, 1965.

[3] Stephen L. Passman, Jace W. Nunziato and Edward K. Walsh, *A Theory of Multiphase Mixtures*, in *Appendix 5C of Rational Thermodynamics*, 2nd edition, Springer-Verlag, New York, 1984.

[4] Ray M. Bowen, *Theory of Mixtures*, in *Continuum Physics*, Vol. III, edited by A.C. Eringen, Academic Press, New York, 1976.

[5] David F. McTigue, Richard C. Givler and Jace W. Nunziato, *Rheological effects of nonuniform particle distributions in dilute suspensions*, Journal of Rheology, 30, 5 (1986), pp. 1053–1076.

[6] Stephen L. Passman, *Forces on the Solid Constituent in a Multiphase Flow*, Journal of Rheology, 30, 5 (1986), pp. 1077–1083.

[7] J. D. Ramshaw and J. A. Trapp, *Characteristics, stability, and short wavelength phenomena in two-phase flow equation systems*, Nuclear Science and Engineering 66 (1978), pp. 93–102.

---

[7] In another paper in this volume, J. T. Jenkins and D.F. McTigue present a similar set of ideas for a very concentrated suspension. The kinetic theory is worked out in considerable detail. D. F. McTigue also has presented a similar set of ideas in an unpublished lecture at the meeting of the American Geophysical Union held in December, 1988.

[8] C. PAUCHON AND S. BANNERJEE, *Interphase momentum interaction effects in the averaged multifield model, Part I: Void propagation in bubbly flows*, International Journal of Multiphase Flow, 12 (1986), pp. 555–573.

[9] C.J.S. PETRIE, *Elongational Flows*, Pitman, London, 1979.

[10] C. TRUESDELL AND W. NOLL, *The Non-Linear Field Theories of Mechanics, Handbuch der Physik III/3*, Springer-Verlag, 1965. A modern discussion is given in P. Podio-Guidugli and M. Gurtin, *The Thermodynamics of Constrained Materials*, Archive for Rational Mechanics and Analysis, 51, (1973), pp. 192–208.

[11] C. TRUESDELL, *A First Course in Rational Continuum Mechanics, Volume 1*, Academic Press, New York, 1977.

[12] J.O. HIRSCHFELDER, C. F. CURTISS AND R. BYRON BIRD, *Molecular Theory of Gases and Liquids*, Wiley, 1954.

[13] J. C. MAXWELL, *On the Dynamical Theory of Gases*, Philosophical Transactions of the Royal Society, London, 157 (1867), pp. 49–88.

[14] C. TRUESDELL, *Sulle Basi di Termomeccanica*, Accademia Nazionale dei Lincei, Rendiconti della Classe di Scienze Fisiche, Mathematiche e Naturali, Series 8, 22 (1957), pp. 33–38, 158–166; (An English translation by the author appears as *On the Foundations of Mechanics and Energetics, Continuum Mechanics II*, Gordon & Breach, New York, 1965, 292–305.

COMPUTATIONS OF GRANULAR FLOW IN A HOPPER

E. BRUCE PITMAN*

Abstract. The flow of granular material in a hopper is a common industrial problem, but it is a problem without a good solution. Classical theories treat the material as an incompressible continuum in steady plastic yield; such theories cannot explain experimentally observed dynamics and dilantcy. Investigation of dynamic theories which include density variation is just beginning. We review the classical theory of granular flow in bins and present some of the recent developments on compressible flows. We borrow ideas from computational fluid dynamics in order to develop a method for the numerical simulation of compressible hopper flow.

§1 INTRODUCTION

Many industrial settings require the storage of granular materials in bins, and the subsequent removal of those materials. One industrial example is the storage of coal at electric generating plants; another example is the storage of pharmaceuticals. Different requirements drive the design of storage vessels in these two examples. In the first case, bunkers are built to hold large quantities of coal; the major design aim is the structural integrity of the bin, with reasonable flow characteristics during emptying. In the second case loads on the bin are small and the hopper is built to ensure excellent flow capabilities with little particle segregation during discharge.

The coal and pharmaceutical examples illustrate the range of questions which must be addressed in the design of storage vessels. Theories of material deformation must incorporate sufficient information about particle-particle and particle-bin slip and rolling to allow prediction of stresses and velocities throughout the material and at the hopper walls. Many of the classical ideas of granular flow are static or quasi-static theories based on simple sliding friction. More complete theories are needed in order to examine dynamic phenomena, especially the propagation of dilantcy waves (see Behringer's report [2] in these proceedings). Related work is underway in soil and rock mechanics, where dynamic deformation is an important ingredient in the larger study of soil-structure stability.

In this paper, we examine granular flow in hoppers. We begin by introducing some of the classical ideas of granular flow in Section 2, highlighting the successes of the theory and its limitations. In Section 3 we introduce the Critical State Theory and discuss the results of simple mathematical analysis of the theory. Finally in Section 4 we use the Critical State Theory to develop a numerical scheme for the simulation of a one dimensional hopper flow.

---

*Institute for Mathematics and its Applications, University of Minnesota, Minneapolis, Minnesota 55455 and Department of Mathematics, State University of New York, 106 Diefendorf Hall, Buffalo, NY 14214–3093. This research is supported by the Air Force Office of Scientific Research under grant AFSOR 88-0182, and the NSF and ONR through grants to the Institute for Mathematics and its Applications and the Minnesota Supercomputer Institute.

§2 CLASSICAL THEORIES OF GRANULAR FLOW IN A HOPPER

In this section, we discuss some of the classical ideas of hopper flows. By the term "classical" we do not mean "old". Rather, these theories consider the granular material as an incompressible continuum which deforms according to simple plasticity rules. In addition, these theories usually assume quasi-static flow; time derivatives (and often all inertial terms) are dropped from the conservation equations. In the first subsection, we detail the basics assumptions and results of one such classical theory; in the second subsection we outline some of the difficulties of this theory.

(a) Incompressible Mohr-Coulomb Theory. We outline a simple theory for the incompressible flow of a granular material; the ideas originated with the work of Coulomb in the eighteenth century and were extended by Mohr at the turn of this century. In this discussion we follow Jenike [6]. For ease of exposition, we formulate the equations in two Cartesian dimensions; later we shall change to a spherical/polar coordinate system appropriate for hopper flows. The fundamental equations of motion are the balance laws for mass and linear momentum. In fixed Eulerian coordinates these equations take the form:

$$d_t\rho + \rho\partial_j v_j = 0 \tag{2.1a}$$

$$\rho d_t v_i + \partial_j T_{ij} = b_i. \tag{2.1b}$$

Here ρ is the bulk density of the material at the point $x = (x_1, x_2)$ and time t, $v = (v_1, v_2)$ is the velocity, T_{ij} is the Cauchy stress tensor, and b_i are body forces (e.g. gravity). We use d_t to denote the convective derivative $d_t = \partial_t + v_j\partial_j$, and employ the summation convention, where the indices $i, j = 1, 2$. We fix a sign convention by regarding compressive stresses as positive.

Equations (2.1) are three equations for the 6 unknowns ρ, v_i, T_{ij} (Note: Conservation of angular momentum implies T is symmetric). In order to close the system, we must make assumptions relating the field variables. The first simplification is to drop all inertial terms; that is, $d_t \equiv 0$. We reduce the number of dependent variables by assuming the material is incompressible; then ρ is a constant and the continuity equation becomes $\nabla \cdot v = 0$. This still leaves more unknowns than equations. We now postulate constitutive laws relating T to v. These relations are: (i) the Yield Condition; (ii) the Flow Rule.

The Yield Condition is an algebraic relation among the stresses. Let us write the eigenvalues of T as $\sigma_1 \geq \sigma_2$, and define the mean and shear stress, respectively,

$$\sigma = \frac{\sigma_1 + \sigma_2}{2} \qquad \tau = \frac{\sigma_1 - \sigma_2}{2}.$$

We assume that the material is at yield, which implies σ and τ are proportional

$$\tau = sin(\delta)\sigma; \tag{2.2}$$

here δ is the angle of internal friction, a material constant.

The Flow Rule is a relation between the stress and the strain-rate tensors. Here the strain-rate, V, is defined by $V_{ij} = -\frac{1}{2}(\partial_i v_j + \partial_j v_i)$. We assume that the eigenvectors of the stress and strain-rate are co-linear. In order to express this relation in a convenient form, we follow [6] and perform a non-linear change of stress variables to obtain

$$T = \sigma \begin{pmatrix} 1 & 0 \\ 0 & 1 \end{pmatrix} + \sigma sin(\delta) \begin{pmatrix} \cos(2\psi) & \sin(2\psi) \\ \sin(2\psi) & -\cos(2\psi) \end{pmatrix}. \tag{2.3}$$

Here, ψ is the angle between the of major principal stress axis (i.e., the eigenvector associated with the eigenvalue σ_1) and the x-axis. With this definition of stress, the yield condition 2.2 is automatically satisfied. The flow rule takes the form

$$\sin(2\psi)(\partial_x v_1 - \partial_y v_2) = \cos(2\psi)(\partial_y v_1 + \partial_x v_2) \tag{2.4}$$

The equations 2.1, 2.3, 2.4 constitute a closed system of four time independent hyperbolic PDE for the unknowns σ, ψ, v_1, v_2.

(b) The Radial Field. A change of independent variables will convert 2.1, 2.4 into the polar coordinate system as shown in Figure 2.1. We assume symmetry in the "hoop" direction, leaving r and θ as dependent variables, with u the r-velocity and v the θ-velocity. Jenike [6] found a similarity solution to these equations in an infinite hopper, the Radial Field:

$$\sigma = rs(\theta) \quad \psi = g(\theta) \quad u = -r^{-2}w(\theta) \quad v = 0. \tag{2.5}$$

The Radial Field solution does not determine the absolute magnitude of the velocity, only its distribution across the hopper. The magnitude is set by conditions at the hopper exit. The Radial Field continues to serve as a basis for hopper design both in the U.S. and abroad.

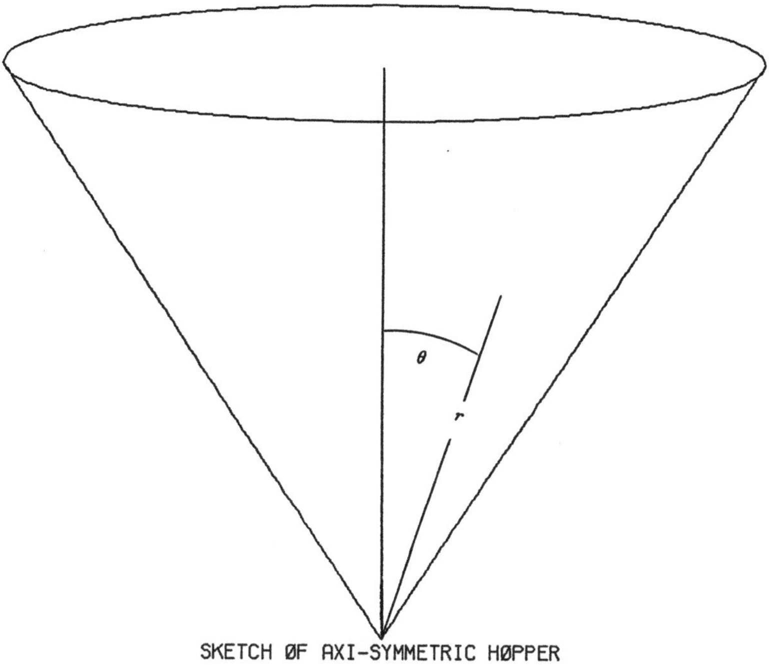

FIGURE 2.1

In order to account for the finite size of a real hopper and to include inertial effects, investigators use the Radial Field as the first term in an expansion of the complete solution [3, 10, 12]. The validity of such expansions is called into question by the results of [13], which investigated the linear stability of the Radial Field. Let us explain the technique employed. In the steady state, the $r \to 0$-direction may be considered the "forward in time" direction. This interpretation transforms steady hopper flows into a one space dimension plus "time" initial-boundary value problem for the stresses and velocities. Flow is linearized about the Radial Field and one asks what happens, as flow proceeds down the hopper (i.e. in "future time"), to a small fluctuation from the Radial value of some dependent variable. The answer is that, depending on material parameters, some flows are unstable - the fluctuations may grow.

In addition to the stability results of [13], experimental work casts doubt on the assumptions made in deriving the Mohr-Coulomb theory. Drescher [4] shows the alignment assumption 2.4 fails in many circumstances. More recent experiments [2, 11] show dynamic density variations and voidage waves which are inconsistent with the assumption of steady incompressible flow. In the next section we outline a theory of granular flow which incorporates the density as a field variable.

§3 THE CRITICAL STATE THEORY

In the late 1950's, Roscoe [17] and colleagues began investigations which led to the development of the Critical State Theory of Soil Mechanics. Recent efforts [5] have adapted these ideas to flowing granular materials. In the first sub-section, we outline this theory; in sub-section (b), we examine the equations for hopper flows.

(a) Critical State Theory for Granular Materials. We again begin the exposition by considering equations 2.1 in two Cartesian dimensions. In the Critical State Theory we assume a plastic yield condition of the form:

$$\Phi(T,\rho) = 0. \tag{3.1}$$

For fixed ρ, this equation defines a closed surface in stress space. In the interior of this surface material would behave elastically; we make the rigid-plastic assumption which means deformation may occur only for stresses on the surface.

A precise form of this yield condition will be given in the next sub-section. We shall require that Φ be a twice continuously differentiable function in the stress variables and once differentiable in the density. Physical arguments show that Φ must be monotonically increasing in ρ; we assume that the yield surfaces scale in a self-similar way as the density is varied, generating a nested sequence of yield loci. For ρ fixed, Φ is assumed to be a strictly convex function of the stresses.

We adopt an Associated Flow Rule given by

$$\frac{\partial \Phi}{\partial T_{ij}} = \mu V_{ij}, \quad \mu \geq 0. \tag{3.2}$$

In 3.2, μ is not specified a priori, but must be determined as part of the solution procedure. That is, equation 3.2 may be viewed as a prescription for finding the ratios of components of the strain-rate tensor, but not the norm of the tensor. This norm, or, equivalently, the magnitude of the velocities, is set by boundary conditions.

For fixed ρ, there is a unique stress point at which 3.2 predicts flow without a change in density; this point is called the Critical State. Because of the self-similarity of the yield surfaces, the critical states for different densities all lie along a common line. This is the line of constant proportionality (equation 2.2) in the incompressible theory. On one side of this line of critical states, flow is accompanied by consolidation; on the other side, by expansion.

In two dimensions, 3.2 consists of three equations for the independent entries of V, and 3.1 provides one algebraic relationship. Coupled with equations 2.1, we have seven equations for the seven unknowns (ρ, v, T, μ). Assuming the material at yield, we may solve 3.1, 3.2 for the stress and μ as functions of the density and strain-rates and write

$$T = T(V,\rho). \tag{3.3}$$

(Remark: T, when considered as a function of the velocities, is homogeneous of degree zero; see equation 3.8 below.) Now substitute (3.3) into (2.1) to obtain a system of evolution equations in the variables ρ, v:

$$d_t\rho + \rho\partial_j v_j = 0 \tag{3.4a}$$

$$\rho d_t v_i + (\frac{\partial T_{ij}}{\partial \rho})\partial_j \rho - (\frac{\partial T_{ij}}{\partial V_{kl}})\partial_j\partial_k v_l = b_i. \tag{3.4b}$$

Equations (3.4) resemble the compressible Navier-Stokes equations. However, the nature of the viscous dissipation is very different for the two systems. In particular, the symbol matrix associated to $\frac{\partial T}{\partial V}$ is only semi-definite for flow in two dimensions; it vanishes along two distinct lines in Fourier Transform space. This degeneracy can lead to ill posedness (i.e. infinite growth of a Fourier mode for the linearized equations). $\frac{\partial T}{\partial V}$ is negative definite for three dimensional flows. See [14, 19] for details. Other studies have described possible instabilities (i.e. finite positive growth of a Fourier mode) in the Critical State equations [15,20,21]. The important result for us is that, while hopper flows should remain well posed for all time, instabilities may develop.

(b) Equations for Hopper Flows. In this section we develop the critical state equations for hopper flows. We shall consider flow in an axi-symmetric hopper with dependence only on the r-direction; let u be the velocity in this direction.

For simplicity, we chose as a yield surface an ellipsoid of revolution, whose cross-section (in any plane which contains the hydrostatic axis) is given by

$$\frac{(\sigma - P)^2}{c^2P^2} + \frac{\tau^2}{s^2P^2} = 1. \tag{3.5}$$

Here σ is the mean stress and τ is the stress coordinate which lies in the chosen plane and is orthogonal to σ. We have introduced $P = \rho^{\frac{1}{\beta}}$, where $\beta \sim 10^{-1} - 10^{-2}$ is a measure of the compressibility of the material; $s = sin(\delta)$ as defined above, and c is a measure of the cohesion of the material. We shall consider cohesionless materials, defined by $c = 1$.

The assumption of one dimensionality simplifies the computation of V. The three non-zero strain rates are

$$\epsilon_{rr} = \partial_r u \qquad \epsilon_{\theta\theta} = \epsilon_{\phi\phi} = u/r. \tag{3.6}$$

It is convenient to divide V into spherical and deviatoric parts:

$$V = -\frac{1}{3}\left[(\partial_r u + 2\frac{u}{r})\begin{pmatrix}1&0&0\\0&1&0\\0&0&1\end{pmatrix}(\partial_r u - \frac{u}{r})\begin{pmatrix}2&0&0\\0&-1&0\\0&0&-1\end{pmatrix}\right] \tag{3.7}$$

Let $\alpha = \sqrt{2/3}\, tr(V)/|dev(V)|$, where $tr(V)$ is the trace of the matrix V and $|dev(V)|$ is the Euclidean norm of its deviator. Solving 3.2, 3.5, we find:

$$T = P\left[(1 + \frac{\alpha c^2}{\sqrt{s^2 + c^2\alpha^2}})I(\frac{s^2}{\sqrt{s^2 + c^2\alpha^2}})\frac{dev(V)}{|dev(V)|}\right] \tag{3.8}$$

where I is the 3×3 identity matrix.

Under the assumptions stated, the continuity equation becomes

$$\partial_t \rho + \partial_r(\rho u) = -2\frac{\rho u}{r} \tag{3.9a}$$

while the r-momentum equation is

$$\partial_t(\rho u) + \partial_r(\rho u^2 + T_{rr}) = 2\frac{T_{rr} - T_{\theta\theta}}{r} - 2\frac{\rho u^2}{r} - \rho g. \tag{3.9b}$$

In 3.9b, we have taken the gravitational body force to act in the radial direction (see [16,18]).

The steady state version of equations 3.9 possesses a similarity solution, the analogue of the Radial Field:

$$\rho = \rho_0 r^{\beta/(1-\beta)} \qquad \sigma = \sigma_0 r^{1/(1-\beta)} \qquad u = -u_0 r^{(-2+\beta)/(1-\beta)} \tag{3.10}$$

where ρ_0 and σ_0 are related through the momentum equation 3.9b and u_0 is determined by boundary conditions at the exit.

§4 NUMERICAL METHODS

When considering methods for the numerical solution of equations 3.9, two factors need to be kept in mind. First is the possible instability (or, in more general flow configurations, ill posedness) alluded to above. Additional dissipation terms may be required, and the integration techniques to be employed must not numerically excite high frequency modes which would otherwise remain quiescent. The second fact to recognize is that flow proceeds on two very different time scales. The momentum equations act on a fast time $\tau_1 \sim U/g$ while the continuity equation evolves on a slow scale $\tau_2 \sim \beta L/U$. In typical hopper flows, and away from the bottom of the bin, the characteristic velocity $U \sim 10^{-1}$ m/sec and the characteristic macroscopic length $L \sim 10$ m, so $\tau_1 = O(10^{-2})$ and $\tau_2 = O(10^0)$. This scaling suggests an implicit method be used for solving 3.9. (Remark: One may exploit the separation of scales and solve only on the slow time, assuming dynamics on the fast scale equilibrate completely on times defined by the slow scale. This is the strategy of the quasi-dynamic approximation introduced in [15, 20, 21].)

One possible regularization of 3.9 is suggested by a careful consideration of the dynamics occurring in a region of high shearing. In such a region there is a large transfer of momentum over distances on the order of an individual grain. Jenkins [7], Savage [8], and colleagues have developed a theory for granular flow at high shear rates based on an extension of the standard kinetic theory to slightly inelastic particles. Incorporating the Enskog dense gas correction, the continuum equations for this granular fluid are the Navier-Stokes equations for a compressible, heat conducting fluid, with the coefficients of viscosity and heat conductivity depending on the field variables. The "heat" is not molecular energy, but the so-called "granular temperature" measuring a particle's deviation from the mean motion of its neighbors. There is one non-classical term in the theory, a heat sink which represents the energy loss due to inelastic collisions.

Our idea, following [9], is to add the stresses of the kinetic theory to the critical state stresses. Let the stresses derived from the kinetic theory be denoted T^k, and those from the critical state theory be T^f and define the total stress to be

$$T = T^f + T^k. \tag{4.1}$$

In the computations considered here, we fix the granular temperature to be constant; the kinetic terms then provide a regularization to the momentum equation without introducing any extra equations. Justification for this approximation needs to be investigated. We comment that, when appropriately non-dimensionalized, contributions from the kinetic theory are scaled by a factor which is $O(\eta)$, where η is the ratio of a grain diameter to the characteristic length L.

The numerical scheme we employ comes from the literature on computational fluid dynamics. We use a Beam-Warming scheme [1], as corrected by Yee and Harten [22, 23]. In the version employed here (see [22]), Beam-Warming is a second order accurate (in space and time), implicit, three level scheme, requiring a block tri-diagonal matrix inversion at each time step. Nonlinearities in the implicit time stepping are removed by a Taylor expansion. In our implementation, a number of terms in this linearization are proportional to β, which is presumed small; we simplify the computations by dropping all such terms. (In the context of the compressible Navier-Stokes equations, a similar simplification is to assume the viscosity and heat conductivity are independent of the field variables.) Harten and Yee [23] introduce TVD-type corrections in order to prevent numerical oscillations which may arise from a blind application of centered differencing.

(Remark: Although in this paper we are only solving a one dimensional problem, our ultimate aim is to compute a fully three dimensional flow configuration. By using an ADI factorization, Beam-Warming allows for efficient solution of multi-dimensional problems.)

We now proceed to explain the computation of one particular problem. In typical hopper flows, a feeder at the bin exit sets the overall flow rate, i.e. the velocities. Consider flow which has reached steady state, with the magnitude of u at the exit of, say, 0.75. This steady state will be taken as initial data. At the instant $t = 0$, increase the flow rate to 1.0. We wish to detail the transient behavior of the system as it moves toward a new steady state.

Boundary conditions for the problem are as follows. The characteristic length is taken to be the distance from the virtual vertex of the hopper to its physical bottom; this distance is scaled to 1. Along $r = 1$, the velocity u is given. The top of the hopper is taken at $r = 5$, where u and P are prescribed; we assume the top of the bin is continuously charged with new material, whose density and velocity are consistent with equation 3.9. In the figure shown below, we use 61 points in the computational domain, with a constant time step $\Delta t = 0.001$. The angle $\delta = 45$ and the compressibility $\beta = 1/10$. The computations were done on an Apollo work station, Series 4500; a run of 10 scaled seconds (10,000 time steps) took about 15 minutes elapsed time.

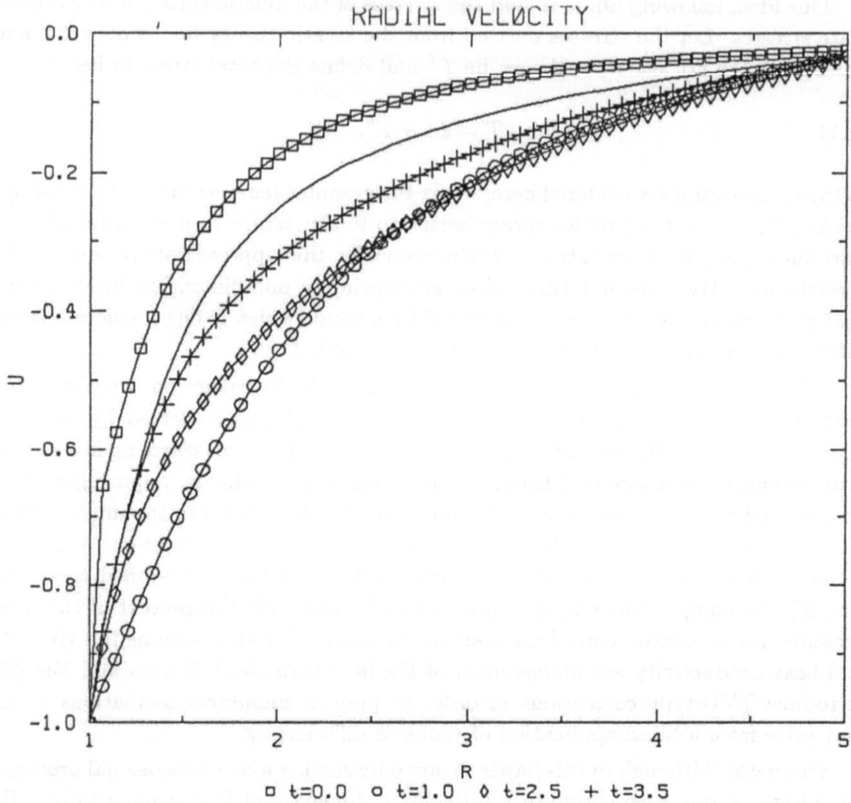

FIGURE 4.1

We show here only the radial velocity profile at various times. The solid line in the figure is the steady-state, reached after approximately 10 seconds. Note how the velocity increases quickly throughout the bin, especially near the bottom. The bottom region then slows down, forcing the system towards its steady state. Keep in mind that all of this variation occurs on the fast time scale. Not shown here are density profiles, which take a much longer time to settle down.

REFERENCES

[1] R. M. Beam and R. F. Warming, *An Implicit Factored Scheme for the Compressible Navier-Stokes Equations*, AIAA Jour., 16 (1978), pp. 393-402.

[2] R. Behringer, these proceedings.

[3] C. Brennan and J. C. Pearce, *Granular Material Flow in Two Dimensional Hoppers*, J. Appl. Mech., 45 (1978), pp. 43-50.

[4] A. Drescher, *An Experimental Investigation of Flow Rules for Granular Materials using Optically Sensitive Glass Particles*, Geotech., 26 (1976), pp. 591-601.

[5] R. Jackson, *Some Mathematical and Physical Aspects of Continuum Models for the Motion of Granular Materials*, in *The Theory of Dispersed Multiphase Flow*, R. Meyer (ed.), Academic Press (1983).

[6] A. W. Jenike, *Steady Gravity Flow of Frictional-Cohesive Solids in Converging Channels*, J. Appl. Mech., 31 (1964), pp. 5-11.

[7] J. T. Jenkins and M. W. Richman, *Grad's 13 Moment System for a Dense Gas of Inelastic Spheres*, Arch. Rat'l. Mech. Anal., 87 (1985), pp. 355-377.

[8] J. T. Jenkins and S. B. Savage, *A Theory for the Rapid Flow of Identical, Smooth, Nearly Elastic Spherical Particles*, J. F. M., 130 (1983), pp. 187-202.

[9] P. C. Johnson and R. Jackson, *Frictional-Collisional Constitutive Relations for Granular Materials*, J. F. M., 176 (1987), pp. 67-93.

[10] K. R. Kaza and R. Jackson, *The Rate of Discharge of Coarse Granular Material from a Wedge Shaped Hopper*, Powder Tech., 33 (1982), pp. 223-242.

[11] R. Michalowski, *Flow of a Granular Material through a Plane Hopper*, Powder Tech., 39 (1984), pp. 29-40.

[12] Z. Mroz and C. Szymanski, *Gravity Flow of a Granular Material in a Converging Channel*, Arch. Mech. Stos., 23 (1971), pp. 897-917.

[13] E. B. Pitman, *The Stability of Granular Flow in Converging Hoppers*, SIAM J. Appl. Math., 48 (1988), pp. 1033-1053.

[14] E. B. Pitman and D. G. Schaeffer, *Stability of Time Dependent Compressible Granular Flow in Two Dimensions*, Comm. Pure Appl. Math., 40 (1987), pp. 421-447.

[15] E. B. Pitman, D. G. Schaeffer and M. Shearer, *Stability in Three Dimensional Critical State Theories of Plasticity*, in preparation.

[16] J. R. Prakash and K. K. Rao, *Steady Compressible Flow of Granular Material through a Wedge Shaped Hopper: The Smooth Wall Radial Gravity Problem*, Chem. Eng. Sci., 43 (1988), pp. 479-494.

[17] K. H. Roscoe, A. N. Schofield and C. P. Wroth, *On the Yielding of Soils*, Geotech., 8 (1958), pp. 22-53.

[18] S. B. Savage, *Mass Flow of Granular Material from Coupled Velocity-Stress Fields*, Brit. J. Appl. Phys., 16 (1965), pp. 1885-1888.

[19] D. G. Schaeffer and E. B. Pitman, *Ill Posedness in Three Dimensional Plastic Flow*, Comm. Pure Appl. Math., 41 (1988), pp. 879-890.

[20] D. G. Schaeffer, M. Shearer and E. B. Pitman, *Instability in Critical State Theories of Granular Flow*, to appear, SIAM J. Appl. Math..

[21] M. Shearer and D. G. Schaeffer, *Foundations of the Quasi-dynamic Approximation in Critical State Plasticity*, preprint.

[22] H. C. Yee, *Linearized Form of Implicit TVD Schemes for the Multidimensional Euler and Navier- Stokes Equations*, Comp. Math. Appl., 12 (1986), pp. 413-432.

[23] H. C. Yee and A. Harten, *Implicit TVD Schemes for Hyperbolic Conservation Laws in Curvilinear Coordinates*, AIAA Jour., 25 (1987), pp. 266-274.

STABILITY OF TWO-PHASE FLOW MODELS

ANDREA PROSPERETTI* AND JAMES V. SATRAPE*

Abstract. A general class of incompressible two-phase flow models containing only algebraic and first-order differential terms is considered. It is shown that the stability of this class of models is independent of the wavenumber of the perturbations. Therefore hyperbolicity is a necessary, although not sufficient, condition for stability. A number of two-phase models available in the literature are examined in the light of the general results. It is found that most of them fail to be stable even in the parameter ranges where they are hyperbolic. Some comments on models with higher-order derivatives are also given.

Key Words. two-phase flow, stability, hyperbolicity

AMS(MOS) subject classification. 76T05

1. Introduction. The modeling of engineering multi-phase flows presents such formidable difficulties that, in spite of at least two decades of concentrated efforts, serious uncertainties still exist. While the structure of some of the terms that must be included in the equations can, to some extent, be guessed, there is no guarantee that this structure is correct nor that all the necessary terms have been included. For the large-concentration cases for which no exact mathematical approach (such as that pioneered, for example, by Batchelor [1]-[3]) is available, a number of researchers have looked to the fundamental principles of Continuum Mechanics [4]-[6] and to the entropy inequality [7] to constrain the form of the equations. While some headway has been made by these approaches, the degree of latitude which still remains in the formulation of the equations is so large that one cannot hope to make substantial progress by relying on experiment alone. Additional constraints are needed to narrow down the range of possibilities. It may be hoped that, once this is achieved, detailed experiments and possibly computations can help resolve those issues still outstanding.

For this reason the last few years have witnessed a renewed interest in the application of models to very simple situations. It is expected that, by examining their predictions for cases that can either be solved exactly or for which one can rely on physical intuition, the strengths and weaknesses of the models can be identified and the presence and nature of some of their terms substantiated [8]. While examination of the basic physics cannot but prove useful, this approach encounters some fundamental limitations in the fact that the limit cases that can be analyzed are highly idealized and it is not obvious that an averaged-equation model is, even

*Department of Mechanical Engineering, Johns Hopkins University, Baltimore, Maryland 21218. Support from the National Science Foundations under grant No. MSM-8607732 is gratefully acknowledged. This study was begun when the senior author was a guest at the Institute of Mathematics and its Applications. Some results were obtained with the aid of MACSYMA, a trademark of Symbolics, Inc.

in principle, applicable to them. There is, therefore, the need to identify further constraints that can be brought to bear on the task of reducing the uncertainties to a small enough number that experiment and computation can help develop a set of equations useful for engineering applications and with a sound physical basis. We propose here that the stability features of the models can be useful for this purpose.

Although multi-phase flows are very unstable on short time and space scales, one would expect their averaged description to be free of these small-scale instabilities. This situation would be similar to the case of single-phase turbulent flow where Reynolds' equations accomplish the same task. The instabilities remaining in the equations should correspond to major transitions in the nature of the flow which are important even in an average sense. For example, one may expect that the transition between bubbly and slug flow, or slug and annular flow, might appear mathematically as an instability of averaged equations describing bubbly flow or slug flow respectively. With the exception of such cases, one would expect a "good" set of averaged equations to be stable.

The question of stability is of course related to the much-debated issue of hyperbolicity of the models. It will be shown in this paper that, surprisingly, statements of very general validity can be formulated on the basis of gross features of the mathematical structure of the models, such as the order of the derivatives appearing in them. Part of this work has been reported earlier on the basis of an analysis which, while useful to simplify the algebra, made the results perhaps less transparent than would be desirable [9],[10]. For this reason in this paper we rederive those results following a conceptually more straightforward method. Furthermore, we examine in their light a number of models that have been proposed in the literature.

2. A general class of two-phase models. We consider a class of one-dimensional, first-order, two-phase models for the description of the flow of two individually incompressible phases that can be described by conservation of mass and momentum equations of the general form

$$\frac{\partial \alpha_G}{\partial t} + \frac{\partial \alpha_G V_G}{\partial x} = 0, \tag{2.1}$$

$$\frac{\partial \alpha_L}{\partial t} + \frac{\partial \alpha_L V_L}{\partial x} = 0, \tag{2.2}$$

$$\begin{aligned} \rho_G \left[\frac{\partial (\alpha_G V_G)}{\partial t} + \frac{\partial}{\partial x}(\alpha_G V_G^2) \right] \; + \; \alpha_G \frac{\partial p}{\partial x} &= \sum_{j=G,L} \left(h_{Gj} \frac{\partial V_j}{\partial t} + k_{Gj} \frac{\partial V_j}{\partial x} \right) + \\ &+ \; m_G \frac{\partial \alpha_G}{\partial t} + n_G \frac{\partial \alpha_G}{\partial x} + \rho_G \alpha_G A_G, \end{aligned} \tag{2.3}$$

$$\begin{aligned} \rho_L \left[\frac{\partial (\alpha_L V_L)}{\partial t} + \frac{\partial}{\partial x}(\alpha_L V_L^2) \right] \; + \; \alpha_L \frac{\partial p}{\partial x} &= \sum_{j=G,L} \left(h_{Lj} \frac{\partial V_j}{\partial t} + k_{Lj} \frac{\partial V_j}{\partial x} \right) + \\ &+ \; m_L \frac{\partial \alpha_L}{\partial t} + n_L \frac{\partial \alpha_L}{\partial x} + \rho_L \alpha_L A_L. \end{aligned} \tag{2.4}$$

Here, the indices L and G denote the phases (although we do not have liquid and gaseous phases specifically in mind) and ρ, V, and α indicate density, velocity, and volume fraction respectively, with

$$\alpha_G + \alpha_L = 1. \tag{2.5}$$

At this stage we only need to assume that the quantities A_i, h_{ij}, k_{ij}, m_i, and n_i $(i, j = G, L)$ are functions of the flow variables $V_{L,G}$, $\alpha_{L,G}$, and possibly p. Since we shall be examining the stability of steady uniform flows for which the pressure gradient is typically a constant, the appearance of p in the list of arguments of these functions would result in a system of variational equations with non-constant coefficients. This can be allowed, but then our stability analysis, which is carried out by the usual Fourier method, will be approximate rather than exact and will only be reliable for wavelengths small compared with the characteristic length for the variation of p. If the coefficients of (2.3) and (2.4) do not depend on p, on the other hand, our results are exact in the framework of a linearized stability analysis. Derivatives appear linearly in the above equations, which is sufficient for our purposes. Indeed, any nonlinear term in the derivatives would disappear upon linearization about a steady uniform flow and therefore would not contribute to the variational equations that we shall analyze.

The presence of the terms $A_{G,L}$ in Eqs. (2.3) and (2.4) should be explicitly noted. These terms do not contain derivatives and introduce in the model effects such as body forces and interphase and structure drag.

Although only one pressure appears in the formulation, this does not overly restrict our class of models. Indeed, if the model contains two distinct pressures p_G and p_L, it could still be reduced to the form (2.3) and (2.4) provided that a relation of the form

$$p_L - p_G = F(V_G, V_L, \alpha_G, \alpha_L), \tag{2.6}$$

exists. In this case one could simply use $p = p_L$ in both equations, with the correct pressure in the G-phase accounted for by the function F, or one could define

$$p = \frac{1}{2}(p_G + p_L),$$

upon which, for example,

$$\begin{aligned} p_G &= \frac{1}{2}(p_G + p_L) - \frac{1}{2}(p_L - p_G) \\ &= p - \frac{1}{2}F(V_G, V_L, \alpha_G, \alpha_L), \end{aligned} \tag{2.7}$$

and similarly for p_L. Either one of these procedures would lead to the appearance, at least formally, of only one pressure. We shall encounter two such examples in the following.

In earlier papers devoted to this class of models [9],[10], before carrying out the stability analysis, the pressure was eliminated and use was made of the volume velocity of the flow U defined by

$$U = \alpha_L V_L + \alpha_G V_G, \tag{2.8}$$

to reduce the system to two equations rather than four. This procedure was convenient algebraically but its unconventional nature tended to obscure the meaning of the results. For this reason, in the present paper, we deal with Eqs. (2.1) to (2.4) directly.

3. Steady flow. In the steady situation the two continuity equations imply the constancy in space of the volume fluxes $\mathcal{Q}_i = \alpha_i V_i$ of the individual phases. Solving (2.5) and (2.8) for the volume fractions, we thus find

$$V_G = \frac{\mathcal{Q}_G V_L}{V_L - \mathcal{Q}_L}. \tag{3.1}$$

In a "reduced" phase space (V_G, V_L) for the system (2.1) to (2.4), this relation represents a family of hyperbolas corresponding to different values of the phase volume fluxes [9]. Since in the case of a steady and uniform flow all derivatives vanish except $\partial p/\partial x$, Eqs. (2.3) and (2.4) give

$$\frac{\partial p}{\partial x} = \rho_L A_L = \rho_G A_G. \tag{3.2}$$

Hence, in the phase space (V_G, V_L), all states of steady and uniform flows are represented by points on this line which, mathematically, is the locus of critical points of the system. From the theory of differential equations it follows that all steady flows will evolve along the hyperbolas (3.1) towards the (stable) intersections with this line of critical points. If no stable intersections exist, steady flows must follow the hyperbolas to their asymptotes, which are flows in which one of the phases has an infinite velocity. Evidently this is impossible, and therefore stable states of steady uniform flows must exist. This remark motivates our study of the stability of such flows in the next section.

4. Linear stability. A number of studies addressing the stability character of two-phase flow models is available in the literature (see e.g., [11]-[13]). Our objective here is not to add another such analysis, but rather to show that surprisingly general conclusions can be arrived at solely on the basis of Eqs. (2.1) to (2.4) without any need to be specific about the detailed form of the model nor the physics that it incorporates. While physical content will eventually determine the success of any model, we believe that our results are nevertheless interesting because they illustrate general features intrinsic in the mathematical structure of the equations.

Consider a basic unperturbed state of steady, uniform flow with velocities $\bar{V}_{G,L}$, pressure $\bar{p}$, and volume fractions $\bar{\alpha}_{G,L}$ satisfying (3.2). Perturb this state by adding small disturbances, denoted by a prime, so that

$$V_{G,L} = \bar{V}_{G,L} + v'_{G,L}, \quad p = \bar{p} + p', \quad \alpha_{G,L} = \bar{\alpha}_{G,L} + \alpha'_{G,L}.$$

Substitution of these expressions into the governing equations, linearization, and use of (3.2) lead (upon dropping the overbars for convenience) to

$$\frac{\partial \alpha'_G}{\partial t} + V_G \frac{\partial \alpha'_G}{\partial x} + \alpha_G \frac{\partial v'_G}{\partial x} = 0, \tag{4.1}$$

$$\frac{\partial \alpha_L'}{\partial t} + V_L \frac{\partial \alpha_L'}{\partial x} + \alpha_L \frac{\partial v_L'}{\partial x} = 0, \tag{4.2}$$

$$\begin{aligned}(\rho_G V_G - m_G)\frac{\partial \alpha_G'}{\partial t} &+ (\rho_G V_G^2 - n_G)\frac{\partial \alpha_G'}{\partial x} - \rho_G \alpha_G \frac{\partial A_G}{\partial \alpha_G}\alpha_G' + \\ &+ (\rho_G \alpha_G - h_{GG})\frac{\partial v_G'}{\partial t} + (2\rho_G \alpha_G V_G - k_{GG})\frac{\partial v_G'}{\partial x} - \rho_G \alpha_G \frac{\partial A_G}{\partial V_G} v_G' - \\ &- h_{GL}\frac{\partial v_L'}{\partial t} - k_{GL}\frac{\partial v_L'}{\partial x} - \rho_G \alpha_G \frac{\partial A_G}{\partial V_L} v_L' + \alpha_G \frac{\partial p'}{\partial x} = 0,\end{aligned} \tag{4.3}$$

$$\begin{aligned}(\rho_L V_L - m_L)\frac{\partial \alpha_L'}{\partial t} &+ (\rho_L V_L^2 - n_L)\frac{\partial \alpha_L'}{\partial x} - \rho_L \alpha_L \frac{\partial A_L}{\partial \alpha_L}\alpha_L' - h_{LG}\frac{\partial v_G'}{\partial t} - k_{LG}\frac{\partial v_G'}{\partial x} - \\ &- \rho_L \alpha_L \frac{\partial A_L}{\partial V_G} v_G' + (\rho_L \alpha_L - h_{LL})\frac{\partial v_L'}{\partial t} + (2\rho_L \alpha_L V_L - k_{LL})\frac{\partial v_L'}{\partial x} - \\ &- \rho_L \alpha_L \frac{\partial A_L}{\partial V_L} v_L' + \alpha_L \frac{\partial p'}{\partial x} = 0.\end{aligned} \tag{4.4}$$

We assume that the solution to this linearized perturbation problem can be expanded in normal modes in such a way that, for any flow quantity Φ,

$$\Phi(x,t) = \phi \exp(\sigma t + ikx), \tag{4.5}$$

with k the (real) wavenumber, σ the complex frequency, and ϕ the complex amplitude. After (4.5) is substituted into Eqs. (4.1) to (4.5), the resultant system of four algebraic equations will have a solution if and only if the determinant of the system is zero. This condition requires

$$A\sigma^2 + (Bik + C)\sigma + (Dk^2 + Eik) = 0, \tag{4.6}$$

where

$$A = \alpha_G^2(\rho_L \alpha_L - h_{LL}) + \alpha_L^2(\rho_G \alpha_G - h_{GG}) + \alpha_G \alpha_L (h_{GL} + h_{LG}), \tag{4.7}$$

$$\begin{aligned} B &= \alpha_G^2 V_L(\rho_L \alpha_L - h_{LL}) + \alpha_L^2 V_G(\rho_G \alpha_G - h_{GG}) + \\ &+ \alpha_G^2(\alpha_L \rho_L V_L - k_{LL}) + \alpha_L^2(\alpha_G \rho_G V_G - k_{GG}) + \\ &+ \alpha_G \alpha_L(\alpha_G m_L + \alpha_L m_G + k_{GL} + k_{LG} + V_L h_{GL} + V_G h_{LG}),\end{aligned} \tag{4.8}$$

$$C = -\alpha_L^2 \alpha_G \left(\rho_G \frac{\partial A_G}{\partial V_G} - \rho_L \frac{\partial A_L}{\partial V_G}\right) - \alpha_L \alpha_G^2 \left(\rho_L \frac{\partial A_L}{\partial V_L} - \rho_G \frac{\partial A_G}{\partial V_L}\right), \tag{4.9}$$

$$\begin{aligned} D &= -\alpha_G^2 V_L(\alpha_L \rho_L V_L - k_{LL}) - \alpha_L^2 V_G(\alpha_G \rho_G V_G - k_{GG}) - \\ &- \alpha_G \alpha_L(V_L k_{GL} + V_G k_{LG} + \alpha_G n_L + \alpha_L n_G).\end{aligned} \tag{4.10}$$

$$E = \alpha_G^2\alpha_L^2\left(\rho_G\frac{\partial A_G}{\partial \alpha_G} + \rho_L\frac{\partial A_L}{\partial \alpha_L}\right) - \alpha_L^2\alpha_G V_G\left(\rho_G\frac{\partial A_G}{\partial V_G} - \rho_L\frac{\partial A_L}{\partial V_G}\right) -$$

$$- \alpha_L\alpha_G^2 V_L\left(\rho_L\frac{\partial A_L}{\partial V_L} - \rho_G\frac{\partial A_G}{\partial V_L}\right), \tag{4.11}$$

Equations (4.7) to (4.11) can readily be shown to be identical with the corresponding ones of the earlier paper [9] upon a change from the independent variables $(V_{G,L}, U, p)$ used there to the set $(V_{G,L}, \alpha_G, p)$ used here. The stability condition for the initial-value problem is evidently

$$\text{Real } \sigma \leq 0. \tag{4.12}$$

The dispersion relation (4.6) simplifies somewhat upon the substitution

$$\sigma = \sigma' - i\frac{Bk}{2A}, \tag{4.13}$$

which evidently does not affect the stability criterion (4.12). With this substitution (4.6) becomes

$$A\sigma'^2 + C\sigma' + \left(\frac{B^2}{4A} + D\right)k^2 + i\left(E - \frac{BC}{2A}\right)k = 0. \tag{4.14}$$

It is readily shown that, for an equation of the form $X^2 + pX + q + ir = 0$, with p, q, r real, the conditions that must be satisfied to ensure $\text{Real}(X) \leq 0$ are $p \geq 0$ and $p^2 q \geq r^2$. For the case of Eq. (4.14) the first one of these conditions has the explicit form

$$\frac{C}{A} \geq 0. \tag{4.15}$$

For the second stability condition, each side of the inequality has a factor of k^2/A^4 that cancels so that one finds

$$\frac{C^2}{4}\left(B^2 + 4AD\right) \geq \left(EA - \frac{BC}{2}\right)^2. \tag{4.16}$$

This is the most remarkable result of this study. We have found that the stability condition of any first-order model of the very broad class (2.1) to (2.4) is independent of wavenumber. All wavelengths are either stable or unstable together, a most unusual situation in stability problems.

The solution to (4.6) is

$$\sigma = -i\frac{Bk}{2A} + \frac{-C \pm \left(C^2 - [(B^2 + 4AD)k^2 + 2i(2AE - BC)k]\right)^{1/2}}{2A}. \tag{4.17}$$

At small wavenumbers this gives, approximately,

$$\sigma = -i\frac{Bk}{2A}, \qquad \sigma = -i\frac{Bk}{2A} - \frac{C}{A}. \tag{4.18}$$

Perturbations corresponding to the first root are seen to propagate without dispersion with the velocity $B/2A$. Perturbations corresponding to the second root propagate with the same velocity but also attenuate in time at the rate C/A independently of position. From these results it appears that the first stability condition (4.15) is necessary to stabilize long waves. For large wavenumbers the corresponding approximation is

$$\sigma = -i\frac{k}{2A}[B + C \pm (B^2 + 4AD)^{1/2}], \tag{4.19}$$

from which propagation without dispersion or attenuation is also apparent in the stable case. As will be shown in the next section, reality of the square root appearing in this expression guarantees hyperbolicity of the model.

5. Characteristics. Discussions about whether complex characteristics should be allowed in multi-phase flow models abound in the literature. Those in favor note that in general mathematical terms the most serious pathology associated with complex characteristics is the destruction of continuous dependence of the solution on the initial data at short spatial scales. By its very nature, an averaged equation is inaccurate at these scales, and therefore this seems to be a non-issue. Furthermore, it is argued that any numerical discretization procedure would introduce a minimum resolvable scale, so that any problem associated with processes at smaller scales are irrelevant in practice. Finally, since instabilities are known to occur in multi-phase systems, one might expect these instabilities to appear mathematically in the form of complex characteristics.

These arguments appear very convincing but are based, in fact, on the *wrong premise* that *complex characteristics only affect short scales.* While this is in general true, we have shown above that, for the broad class of models considered here (which, as will be shown below, encompasses virtually all first-order models available in the literature), stability is independent of wavelength. This result implies that, if the existence of complex characteristics leads to instability at short scales, the same instability will also be present at *all* scales. Hence, a non-hyperbolic model is bound to be unstable and any numerical result to the contrary must be the consequence of an excessively dissipative numerical scheme and, possibly, a slow growth rate of the instability. Further considerations on this point will be found in the last section.

The preceding argument can be checked directly by a calculation of the characteristic directions for the system (2.1) to (2.4). Two characteristics are found to be infinite. This is due to the absence of time derivatives of the pressure and of the volume velocity defined by (2.8). The other two characteristic directions are found to be given by

$$A\lambda^2 - B\lambda - D = 0, \tag{5.1}$$

which will evidently only be real provided that

$$B^2 + 4AD \geq 0. \tag{5.2}$$

It is evident from (4.16) that (5.2) is a *necessary* condition for stability, although it is not sufficient in general. In other words, *solutions can be unstable also with a hyperbolic model.* This remark weakens considerably the argument described above which attempts to relate non-hyperbolicity with physical instabilities.

It is easy to verify from Eqs. (4.9), (4.11), and (4.16) that the stability condition of the system of equations differs from the condition for real characteristics only due to the presence of the drag terms $A_{G,L}$, as expected. Again, Eq. (5.2) is readily seen to coincide with the earlier result of Ref. [9].

6. Some specific models. Many two-phase models have been proposed that, under the assumptions of one-dimensional flow and constant density of each phase, can be put in the general form given in Eqs. (2.1) to (2.4). It is therefore interesting to examine, in the light of the previous considerations, some of these models. The results of this analysis will form the object of the next two sections. Here we present the models that will be studied and show how they can be cast in the previous general form. Since for all of them the continuity equations take the form (2.1) and (2.2), we shall only exhibit explicitly the form of the momentum equations.

The model of Drew and Wood. Drew and Wood [14] consider a class of models which is interesting also because it includes as special cases other models such as that of Pauchon and Banerjee [15]. For this model the momentum equations are written (correcting an apparent sign error in front of the last term of Eq. (6.2) [16]) as

$$
\begin{aligned}
\rho_G\left(\frac{\partial\alpha_G V_G}{\partial t}+\frac{\partial\alpha_G V_G^2}{\partial x}\right) \;&+\; \alpha_G\frac{\partial p}{\partial x}=\rho_L\alpha_G\xi\frac{\partial}{\partial x}(V_G-V_L)^2+\rho_G\alpha_G A_G\\
&+\; \rho_L\alpha_G C_{vm}\left[\left(\frac{\partial V_L}{\partial t}+V_L\frac{\partial V_L}{\partial x}\right)-\left(\frac{\partial V_G}{\partial t}+V_G\frac{\partial V_G}{\partial x}\right)\right],
\end{aligned}
\tag{6.1}
$$

$$
\begin{aligned}
\rho_L\left(\frac{\partial\alpha_L V_L}{\partial t}+\frac{\partial\alpha_L V_L^2}{\partial x}\right) \;&+\; \alpha_L\frac{\partial p}{\partial x}=-\rho_L\xi(V_G-V_L)^2\frac{\partial\alpha_L}{\partial x}-\\
&-\; \rho_L\alpha_G C_{vm}\left[\left(\frac{\partial V_L}{\partial t}+V_L\frac{\partial V_L}{\partial x}\right)-\left(\frac{\partial V_G}{\partial t}+V_G\frac{\partial V_G}{\partial x}\right)\right]+\\
&+\; \rho_L\alpha_L A_L+\rho_L\frac{\partial}{\partial x}\left[\alpha_G\alpha_L\left(a_L+b_L\right)\left(V_G-V_L\right)^2\right].
\end{aligned}
\tag{6.2}
$$

Here the subscripts G and L are used to indicate the disperse and continuous phases respectively. The pressure p is identified with the continuous phase pressure and the disperse phase pressure is given by

$$p_G=p-\xi\rho_L(V_G-V_L)^2,$$

which is a relation of the type (2.6). This form is motivated by a consideration of the average pressure over the surface of a particle in potential flow translating with velocity V_G-V_L with respect to the continuous phase. For instance, for an isolated sphere, the exact potential flow solution gives $\xi=1/4$. This value is the one normally used in this model even in non-dilute conditions. Similarly the form of the terms preceded by C_{vm} is motivated by the results for virtual mass effects in potential flow. The coupling constant C_{vm} itself is considered to be a function of α_G, equal to 1/2 in the dilute limit. Different authors suggest different dependencies of C_{vm} on α_G. Zuber [17] suggests

$$C_{vm} = \frac{1}{2}\left(\frac{1+2\alpha_G}{1-\alpha_G}\right). \tag{6.3}$$

Mokeyev [18] gives empirically

$$C_{vm} = \frac{1}{2} + 2.1\alpha_G.$$

For small α_G van Wijngaarden [19] gives

$$C_{vm} = \frac{1}{2} + 1.39\alpha_G,$$

while Pauchon and Banerjee [15] take $C_{vm} = 1/2$ independent of α_G.

The terms a_L and b_L are included in Eqs. (6.1) and (6.2) to model the effects of turbulent (Reynolds) stresses on the phases. For these quantities Nigmatulin [21] gives the values $a_L = 1/6$ and $b_L = -1/2$ and Garipov [20] gives $a_L = -3/20$ and $b_L = -1/20$. Pauchon and Banerjee [15] take instead $a_L = b_L = 0$.

Equations (6.1) and (6.2) can be cast in the form (2.3) and (2.4) with the definitions

$$\begin{aligned}
A_G &= A_L = m_G = m_L = n_G = 0,\\
h_{GG} &= h_{LL} = -\alpha_G C_{vm}\rho_L,\\
h_{GL} &= h_{LG} = \alpha_G C_{vm}\rho_L,\\
k_{GG} &= \alpha_G\rho_L\left[(2\xi - C_{vm})V_G - 2\xi V_L\right],\\
k_{LL} &= -\alpha_G\rho_L\left(2\alpha_L(a_L+b_L)V_G - \left[2\alpha_L(a_L+b_L) - C_{vm}\right]V_L\right),\\
k_{GL} &= -\alpha_G\rho_L\left[2\xi V_G - (2\xi + C_{vm})V_L\right],\\
k_{LG} &= \alpha_G\rho_L\left(\left[2\alpha_L(a_L+b_L) + C_{vm}\right]V_G - 2\alpha_L(a_L+b_L)V_L\right),\\
n_L &= \rho_L(V_G - V_L)^2\left[-\xi + (a_L+b_L)(\alpha_G - \alpha_L)\right].
\end{aligned}$$

The model of Nigmatulin. A number of models have been proposed which can not be cast in the form proposed by Drew and Wood. In general, this is due to the presence of a non-zero n_G term as defined in Eq. (2.3). One such model is that of Nigmatulin [21] who proposes the momentum equations (in the notation of this paper)

$$\begin{aligned}
\rho_G\left(\frac{\partial\alpha_G V_G}{\partial t} + \frac{\partial\alpha_G V_G^2}{\partial x}\right) &+ \alpha_G\frac{\partial p}{\partial x} = -\frac{\rho_L}{3}\alpha_G\frac{\partial}{\partial x}\left[\alpha_G(V_G - V_L)^2\right]\rho_G\alpha_G A_G +\\
&+ \frac{\rho_L}{2}\alpha_L\alpha_G\left[\left(\frac{\partial V_L}{\partial t} + V_L\frac{\partial V_L}{\partial x}\right) - \left(\frac{\partial V_G}{\partial t} + V_G\frac{\partial V_G}{\partial x}\right)\right],
\end{aligned} \tag{6.4}$$

$$\rho_L \left(\frac{\partial \alpha_L V_L}{\partial t} + \frac{\partial \alpha_L V_L^2}{\partial x} \right) + \alpha_L \frac{\partial p}{\partial x} = -\frac{\rho_L}{3} \alpha_L \frac{\partial}{\partial x} \left[\alpha_G (V_G - V_L)^2 \right] \rho_L \alpha_L A_L -$$

$$(6.5) \qquad - \frac{\rho_L}{2} \alpha_L \alpha_G \left[\left(\frac{\partial V_L}{\partial t} + V_L \frac{\partial V_L}{\partial x} \right) - \left(\frac{\partial V_G}{\partial t} + V_G \frac{\partial V_G}{\partial x} \right) \right] +$$

$$+ \rho_L \frac{\partial}{\partial x} \left[\alpha_G \alpha_L (a_L + b_L)(V_G - V_L)^2 \right],$$

which can be reduced to the general form of Section 2 with the same coefficients as for the Drew and Wood model with $C_{vm} = 1/2$, $\xi = -\alpha_G/3$, $a_L + b_L = -1/3$, and

$$n_G = -\frac{\alpha_G \rho_L}{3} (V_G - V_L)^2.$$

The model of Geurst. More recently, Geurst [22] has used variational techniques to develop an inviscid formulation for the two-phase problem of gas bubbles entrained in a liquid. Reduced to one dimension, Geurst's equations become

$$(6.6) \qquad \rho_G \left(\frac{\partial \alpha_G V_G}{\partial t} + \frac{\partial \alpha_G V_G^2}{\partial x} \right) + \alpha_G \frac{\partial p_G}{\partial x} = -M_L^d + \rho_G \alpha_G A_G,$$

$$\rho_L \left(\frac{\partial \alpha_L V_L}{\partial t} + \frac{\partial \alpha_L V_L^2}{\partial x} \right) + \alpha_L \frac{\partial p_G}{\partial x} = M_L^d + \rho_L \alpha_L A_L -$$

$$(6.7) \qquad - \rho_L \frac{\partial}{\partial x} \left[m (V_G - V_L)^2 \right] - \frac{\rho_L}{2} \frac{\partial}{\partial x} \left[(m + \alpha_L m')(V_G - V_L)^2 \right],$$

where the pressure in the disperse phase G is written as

$$p_G = p - \frac{m + \alpha_L m'}{2\alpha_L} \rho_L (V_G - V_L)^2,$$

which also agrees with the form (2.6). The quantity M_L^d represents mutual interaction forces between the phases (other than steady drag) and in the model is expressed as

$$M_L^d = \rho_L \left(\frac{\partial}{\partial t} \left[m (V_G - V_L) \right] + \frac{\partial}{\partial x} \left[m V_G (V_G - V_L) \right] + m (V_G - V_L) \frac{\partial V_G}{\partial x} \right),$$

where the quantity m represents a virtual mass coefficient for which Geurst gives the explicit form

$$(6.8) \qquad m(\alpha_G) = \frac{1}{2} \hat{m} \alpha_G \alpha_L \left(1 - \frac{\hat{m} + 2}{\hat{m}} \alpha_G \right),$$

with $\hat{m}$ a numerical constant. Upon identification with the general form (2.3) and (2.4) we have

$$\begin{aligned} A_G &= A_L = 0, \\ h_{GG} &= h_{LL} = -\rho_L m, \\ h_{GL} &= h_{LG} = \rho_L m, \end{aligned}$$

$$
\begin{aligned}
k_{GG} &= \rho_L\left[\frac{\alpha_G}{\alpha_L}(m+\alpha_L m')(V_G-V_L)-m(3V_G-2V_L)\right],\\
k_{LL} &= \rho_L m(V_G-2V_L),\\
k_{GL} &= \rho_L\left[-\frac{\alpha_G}{\alpha_L}(m+\alpha_L m')(V_G-V_L)+mV_G\right],\\
k_{LG} &= \rho_L m V_G,\\
m_G &= m_L=-\rho_L m'(V_G-V_L),\\
n_G &= \rho_L(V_G-V_L)\times\\
&\times\left(\frac{\alpha_G}{2}(V_G-V_L)m''-\left[V_G-\frac{1}{2}\frac{\alpha_G}{\alpha_L}(V_G-V_L)\right]m'+\frac{\alpha_G}{\alpha_L}\frac{V_G-V_L}{2\alpha_L}m\right),\\
n_L &= -\rho_L\frac{V_G-V_L}{2}\left[\frac{m}{\alpha_L}(V_G-V_L)+m'(V_G+V_L)\right].
\end{aligned}
$$

We shall now study the characteristics and stability properties of these models.

7. Hyperbolicity of Specific Models. The characteristics of all the previous models are real for $V_G = V_L$ and for $\alpha_G = 0$. As will be mentioned in Section 9, the first result is a consequence of Galilean invariance plus the vanishing of the quantities $m_{G,L}$ (when $V_G = V_L$) appearing in Eqs. (2.3) and (2.4) for all the models considered, while the second one follows from the fact that in this limit all models degenerate to the equations for inviscid incompressible flow. In all other cases, reality of the characteristics depends on the coefficients ξ, C_{vm}, a_L, b_L, the ratio ρ_G/ρ_L, and the concentration of the dispersed phase α_G.

Drew and Wood [14] have analyzed the hyperbolicity of their class of models for $\rho_G/\rho_L \ll 1$ and the cases $C_{vm} = 0, 1/2$; $\xi = 0, 1/4$; and $a_L + b_L = 0, -1/3$.[1] Here we present a number of extensions of this analysis in the form of *hyperbolicity boundaries* in the $(\rho_G/\rho_L, \alpha_G)$ plane. In Fig. 1 we show the effect of changes in the values of the parameters ξ and $a_L + b_L$ and of the form of the added mass coefficient for a base case consisting of Zuber's expression (6.3) for the added mass coefficient, $\xi = 1/4$, and $a_L + b_L = 0$. The region above and to the right of the lines corresponds to complex characteristics, while that below and to the left corresponds to the region of hyperbolicity of the model. It is seen in the first panel where curves corresponding to $a_L + b_L$ = -1/3 and -1/5 are also shown that the effect of this parameter is very small. The effect of ξ is, on the other hand, large and shows the stabilizing influence of this parameter. Variations with α_G of the added mass coefficient are seen to restrict the region of hyperbolicity, with the constant value $C_{vm} = 1/2$ resulting in the largest hyperbolic region. This particular curve coincides with the results of Pauchon and Banerjee [15] for the limit of small ρ_G/ρ_L.

Similar comparisons are shown in Fig. 2 for which the base case is as above except that $a_L + b_L = -1/5$. A larger value of ξ is found to be stabilizing, as before, while a constant C_{vm} adds a hyperbolic region for large α_G, but narrows it as a function of the density ratio. Fig. 3 is analogous, but here the base case has $a_L + b_L = -1/3$. All the trends of the previous two figures are confirmed. In particular, for $\xi = 0$, the hyperbolic region vanishes in all cases.

[1] Their hyperbolicity limit of $\alpha_G = 0.27$ for $C_{vm} = 1/2$, $\xi = 1/4$, and $a_L + b_L = -1/3$ appears to be incorrect. The results for the other cases are, however, in agreement with ours.

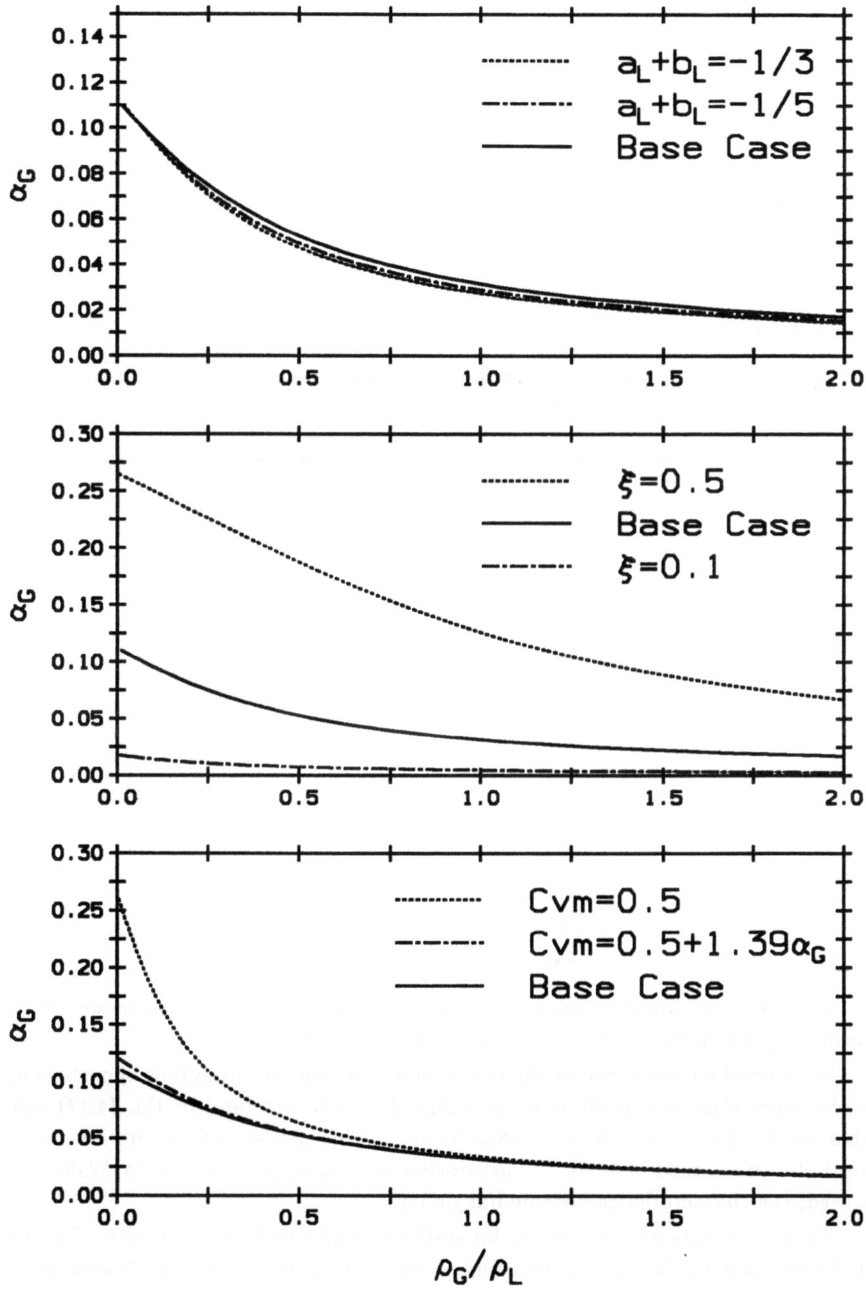

FIG. 1 *Hyperbolicity boundaries for the model of Drew and Wood. The base case is given by* $\xi = 1/4$, $a_L + b_L = 0$, *and Zuber's* $C_{vm} = (1 + 2\,\alpha_G)/(2\,\alpha_L)$. *The region above and to the right of the lines corresponds to complex characteristics, while that below and to the left corresponds to the region of hyperbolicity of the model.*

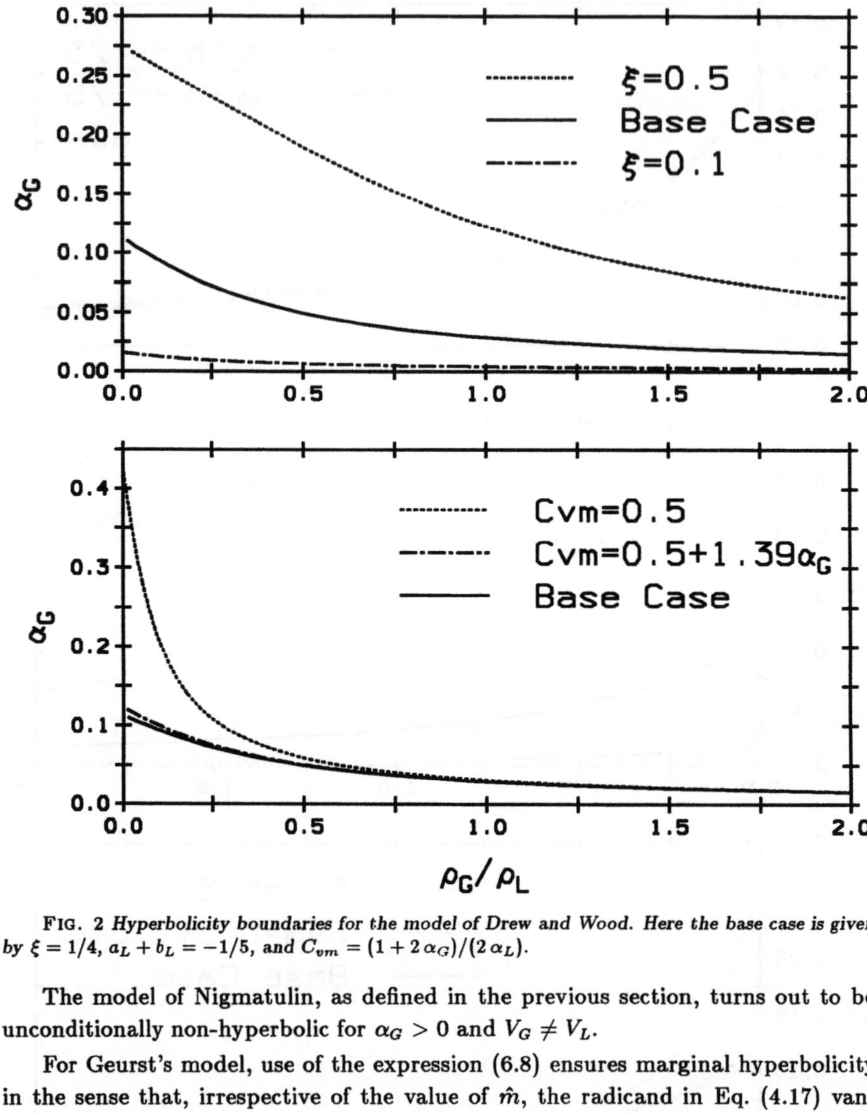

FIG. 2 *Hyperbolicity boundaries for the model of Drew and Wood. Here the base case is given by* $\xi = 1/4$, $a_L + b_L = -1/5$, *and* $C_{vm} = (1 + 2\,\alpha_G)/(2\,\alpha_L)$.

The model of Nigmatulin, as defined in the previous section, turns out to be unconditionally non-hyperbolic for $\alpha_G > 0$ and $V_G \neq V_L$.

For Geurst's model, use of the expression (6.8) ensures marginal hyperbolicity in the sense that, irrespective of the value of $\hat{m}$, the radicand in Eq. (4.17) vanishes resulting in a double real characteristic. This model is therefore effectively parabolic. If, instead of (6.8), the expression $m = \alpha_G \alpha_L / 2$ proposed by Wallis [16] is used, the characteristics become imaginary.

Since hyperbolicity is necessary for stability, in illustrating the stability features of the previous models in the following section, we shall not consider Nigmatulin's model nor Geurst's model with Wallis' added mass term. Most of our comments will be restricted to Drew and Wood's model with Zuber's added mass term.

8. Stability of specific models. In the study of stability we need to be more specific as to the form of the algebraic terms $A_{L,G}$ of the model equations. A

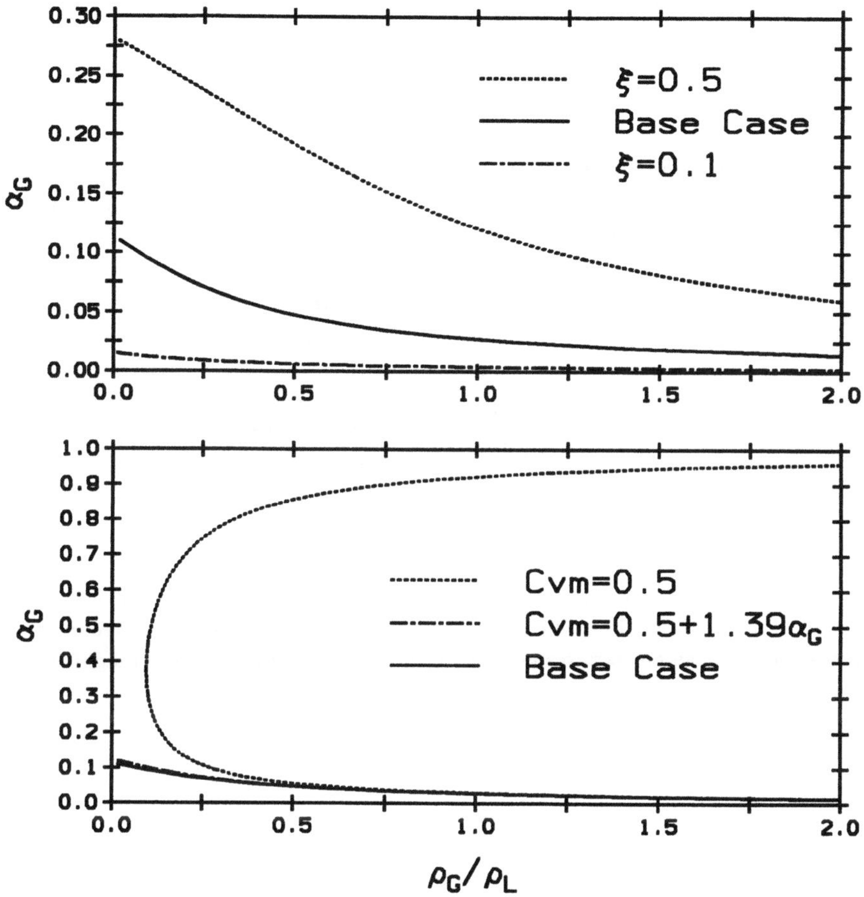

FIG. 3 *Hyperbolicity boundaries for the model of Drew and Wood. Here the base case is given by $\xi = 1/4$, $a_L + b_L = -1/3$, and $C_{vm} = (1 + 2\,\alpha_G)/(2\,\alpha_L)$.*

fairly general form for these terms is

$$A_i = F_i + S_i + g, \tag{8.1}$$

where F_i and S_i represent the interphase and structure drag respectively, and g is the magnitude of the body force in the flow direction x. The action-reaction principle requires that, in steady uniform flow,

$$\rho_G \alpha_G F_G + \rho_L \alpha_L F_L = 0. \tag{8.2}$$

In the sequel, we shall mostly refer to the case of disperse two-phase flows using the subscript G for the disperse phase and the subscript L for the continuous phase. We take the interphase drag to have the form

$$F_G = -K_D |V_G - V_L| (V_G - V_L), \tag{8.3}$$

where K_D depends on α_G and possibly other factors such as particle radius and particle Reynolds number which need not be considered here. F_L is obtained from this equation by use of the relation (8.2). A similar expression for the interphase drag is adopted in many of the existing models and conforms with the general structure proposed by Ishii and Zuber [23] for the interphase drag in disperse flow.

For the structure drag, it is appropriate to assume that $S_L = S_L(\alpha_G, V_L)$ and that, for reasonable concentrations, S_G is negligible. Upon substitution of (8.1) into Eqs. (4.9) and (4.11), assuming that g is constant, and using the relationship (8.2), we find

$$C = -\alpha_G\rho_G\frac{\partial F_G}{\partial V} - \alpha_L\alpha_G^2\rho_L\frac{\partial S_L}{\partial V_L}, \tag{8.4}$$

$$\begin{aligned} E &= \alpha_G\rho_G\left[\alpha_G\alpha_L^2\frac{\partial}{\partial\alpha_G}\left(\frac{F_G}{\alpha_L}\right) - (\alpha_L V_G + \alpha_G V_L)\frac{\partial F_G}{\partial V}\right] - \\ &- \alpha_G\alpha_L\left(\alpha_G\alpha_L\rho_L\frac{\partial S_L}{\partial\alpha_G} + \alpha_G\rho_L V_L\frac{\partial S_L}{\partial V_L}\right), \end{aligned} \tag{8.5}$$

where $V \equiv V_G - V_L$.

Considering the first of the conditions for linear stability given in Eq. (4.16), for all of the models discussed in this paper, A is found to be a positive quantity. If one assumes, as appears reasonable on physical grounds, that the magnitudes of the interphase and structure drags both increase with increasing velocity, C will be a positive quantity and (4.16) will be satisfied.

The second stability condition can only be studied for specific models. To simplify the results, in the following examples we assume that the structure drag is negligible. Upon substitution of (8.3), the quantities C and E become

$$C = 2K_D\alpha_G\rho_G|V_G - V_L|, \tag{8.6}$$

$$E = \alpha_G\rho_G|V_G - V_L|\left[2K_D(\alpha_L V_G + \alpha_G V_L) - \alpha_G(V_G - V_L)(K_D + \alpha_L\frac{\partial K_D}{\partial\alpha_G})\right]. \tag{8.7}$$

Note that both C and E change sign with K_D, so that the second stability condition (4.16) is independent of this sign. The first condition (4.15), however, would be violated unless K_D were positive.

Of the models examined, that of Drew and Wood is the only one where the left-hand side of Eq. (5.2) can be greater than zero. Therefore, it is the only one which may exhibit linear stability when an algebraic drag is added such that the right-hand side of Eq. (4.16) is non-zero. Accordingly, in the following, we will concentrate on the base case given by $\xi = 1/4$, $a_L + b_L = 0$, and $C_{vm} = \frac{1}{2}(1 + 2\alpha_G)/\alpha_L$. We use Zuber's expression for the virtual mass coefficient because it is close to the others but is not explicitly limited to small values of α_G. Noting the relative independence of hyperbolicity on the parameters $a_L + b_L$ as shown in Fig. 1, $a_L + b_L$ is taken to be zero for simplicity.

First consider the case where K_D in Eq. (8.3) is a constant. Using the base case described above and the expressions given in Eq. (8.4) and (8.5), the condition for stability reduces to

$$(8.8)\quad \alpha_G^2\rho_G^2(V_G - V_L)^4K_D^2\left[-9\alpha_G\alpha_L^3\rho - (9\alpha_G^4 + 30\alpha_G^3 - 32\alpha_G^2 + \tfrac{27}{2}\alpha_G - 4)\right] \geq 0.$$

Here, $\rho \equiv \rho_G/\rho_L$. Instability is seen to prevail in all cases other than $\alpha_G = 0$ or $V_G = V_L$. This lack of stability throughout the domain of interest can be shown to exist for all reasonable values of $a_L + b_L$ and for $0 \leq \xi \leq 0.7$.

The assumption that K_D is independent of α_G is probably very unrealistic. In general, one would expect the interphase drag parameter K_D to increase with the volume fraction α_G. Examining the stability conditions given by Eq. (4.16), we find that terms involving $dK_D/d\alpha_G$ appear only on the right-hand side. Consider the expression $(EA - BC/2)$. Using the Drew and Wood coefficients and Eqs. (8.4) and (8.5) we can write

$$(8.9)\qquad EA - \frac{BC}{2} = K_D|V_G - V_L|(V_G - V_L)\rho_G\rho_L\alpha_G^2\left[f - g\frac{K_D'}{K_D}\right],$$

where $K_D' = dK_D/d\alpha_G$ and f and g are given by

$$(8.10)\quad g = \alpha_G\alpha_L\left(C_{vm} + \rho\alpha_L^2 + \alpha_G\alpha_L\right),$$

$$(8.11)\quad f = -3\rho\alpha_G\alpha_L^2 + 2\alpha_L\xi - \alpha_G C_{vm} - 2\alpha_G\alpha_L(a_L + b_L) + \alpha_G\alpha_L(2 - 3\alpha_G).$$

It is evident that $g \geq 0$ everywhere. It can also be shown that, within the range where the characteristics are real, which is necessary for linear stability, $f \geq 0$ for reasonable values of ξ, $a_L + b_L$, and C_{vm}. Since $(EA - BC/2)^2$ appears in the right-hand side of the stability condition and determines a lower bound for stability, the smaller this quantity, the greater the stability region. It is therefore clear that an increase of K_D with α_G has a stabilizing influence provided that gK_D' does not exceed f. To illustrate this point we use a simple power law relation of the form

$$(8.12)\qquad K_D \propto \alpha_G^n \Longrightarrow \frac{K_D'}{K_D} = n\alpha_G^{-1},$$

where $n \geq 0$. With $\xi = 1/4$ and $a_L + b_L = 0$, one finds the model to have some range of stability for $0 < n < 2$. This is shown in Fig. 4 where, in addition to the hyperbolicity boundary (solid line) several stability boundaries corresponding to different values of n are indicated. For $n = 0$ the stable region reduces to $\alpha_G = 0$.

As already mentioned, for Geurst's model with his expression for the added mass terms (6.8), the quantity $B^2 + 4AD$ in the left-hand side of the stability condition (4.16) vanishes. Therefore, unless the drag terms are such as to render identically zero the right-hand side of this equation as well, which is very unlikely, Geurst's model will become linearly unstable in the presence of any non-zero algebraic drag.

9. Constraints on the model equations. The stability and hyperbolicity criteria derived for the general model of Section 2 are very complicated. On the

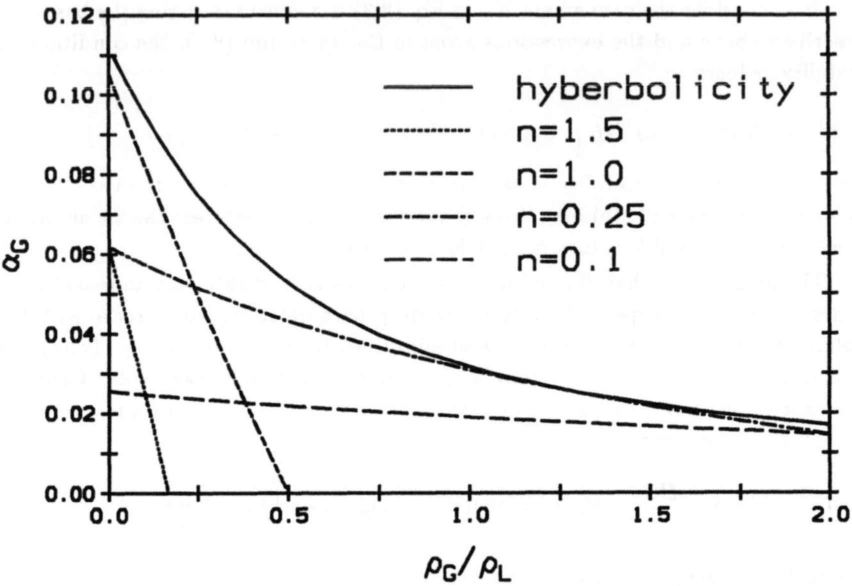

FIG. 4 *Stability boundaries for the model of Drew and Wood with an interfacial drag coefficient proportional to α_G^n. The region above and to the right of the lines is unstable, while that below and to the left is stable.*

other hand, one finds that certain symmetry properties are satisfied in the individual models, and considerable simplifications take place. It is interesting to inquire whether this is a consequence of some basic requirements that all models must satisfy, or is simply an accident due to the particular structure of the models available in the literature. We consider here this point addressing the consequences of the constraints imposed by Galilean invariance and the conservation of total momentum.

Galilean invariance requires that all terms other than the wall drag be independent of the choice of the inertial frame of reference. Consider a transformation such that

$$x = x' + vt, \quad t = t', \quad V_j = V_j' + v. \tag{9.1}$$

Upon substitution of these relations into Eqs. (2.1) to (2.4), the requirement that the form of the equations in both the original and the transformed systems be identical except for the algebraic drag terms leads to

$$\begin{aligned} h_{ij} = h'_{ij}, &\quad k_{ij} - vh_{ij} = k'_{ij}, \\ m_j = m'_j, &\quad n_j - vm_j = n'_j. \end{aligned} \tag{9.2}$$

This result implies immediately that h_{ij} and m_j must either be functions of $V_G - V_L$ or be independent of velocity altogether. Furthermore, it must be possible to express k_{ij} and n_j as

$$k_{ij} = \mu_{ij} V_G + \nu_{ij} V_L, \qquad n_i = \gamma_i V_G + \beta_i V_L, \tag{9.3}$$

where the μ's, ν's ,γ's, and β's are functions of $V_G - V_L$. Combination of (9.2) and (9.3), and the additional requirements that $n_j(V_G, V_L) = n'_j(V'_G, V'_L)$ and $k_{ij}(V_i, V_j) = k'_{ij}(V'_i, V'_j)$, yields

$$h_{ij} = \mu_{ij} + \nu_{ij}, \qquad m_i = \gamma_i + \beta_i. \tag{9.4}$$

These properties are satisfied by all the coefficients of the specific models considered. While it is interesting that this type of structure must exist, it unfortunately does not lead to substantial simplification of the general stability conditions. However, it can be shown from Galilean invariance that the left-hand side of the hyperbolicity condition (5.2) will be identically zero for $V_G = V_L$ provided the coefficients $m_{G,L}$ vanish in this case.

Other relations can be derived by imposing that, upon summation of the two momentum equations, the time derivatives of the velocities only appear in the combination

$$\frac{\partial}{\partial t}\left[\alpha_L \rho_L V_L + \alpha_G \rho_G V_G\right]. \tag{9.5}$$

This requirement is justified by noting that, whatever the specific model considered, this combination must be identified with the total momentum per unit of volume of the mixture. This is a consequence of the definition of the V_j's as center-of-mass velocities of each phase which is implicit in the form of the continuity equations. In this way one finds

$$h_{GL} = -h_{LL}, \qquad h_{LG} = -h_{GG}. \tag{9.6}$$

Again, these relations are satisfied by all the specific models considered above. As before, this constraint does lead to some simplification of the stability conditions, but not to such an extent as to make a substantial difference in their structure or appearance. Accordingly, we shall not quote the corresponding forms.

10. Conclusions. We have shown that some very general statements can be made concerning the stability and hyperbolicity properties of a very broad class of two-phase flow models which includes many specific examples proposed in the literature. This class of models is characterized by the fact that the phases are individually incompressible and only first-order derivatives of the flow quantities appear. The most surprising result is that, in spite of the presence of non-differential "source" terms in the equations, stability of steady uniform flows is independent of the wavenumber of the perturbation. As a consequence, hyperbolicity is necessary, although not sufficient, for stability.

The general results have been applied to a number of specific models. It has been found that most models, even when they are hyperbolic, become unstable upon the addition to the momentum equations of algebraic terms representing, for example, drag forces. Certain regions of stability for a model of the type studied by Drew and Wood [14] have however been identified. Geurst's model, on the other hand, seems to be destabilized by the presence of algebraic terms.

An analysis similar to the present one for models containing higher order derivatives has already been presented in Ref. [10]. The results of that paper show that, as expected, higher derivatives generally introduce a dependence of the stability condition on the wavenumber of the perturbation. In the long wavelength limit, the stability conditions expressed by Eqs. (4.15) and (4.16) above are recovered. For these higher-order models, this result implies instability for long wavelengths unless the "truncated" version of the model (in which only first-order derivatives are retained) is hyperbolic. It is easy to construct examples (e.g., stratified flow) in which this instability is a physical one. In this case the "truncated" model would be non-hyperbolic in certain parameter ranges, but the short-wavelength pathologies usually associated with this feature would (and should, in a good model) be taken care of by the higher-order differential terms. It is only in this framework that non-hyperbolic, first-order models can be considered to be acceptable approximations to a complete model.

In the light of the previous considerations, a lack of hyperbolicity not corresponding to physical instabilities signals some basic problems with a first-order model that cannot be "cured" by the addition of higher-order derivatives. It would therefore seem that such a model cannot be considered in any sense an approximation to a hitherto yet undiscovered "good" model.

REFERENCES

[1] G.K. BATCHELOR, *Sedimentation in a dilute dispersion of spheres*, J. Fluid Mech., 52 (1972), pp. 245-268.

[2] G.K. BATCHELOR, *Sedimentation in a dilute polydisperse system of interacting spheres. Part 1. General theory*, J. Fluid Mech., 119 (1982), pp. 379-408.

[3] G.K. BATCHELOR, *Diffusion in a dilute polydisperse system of interacting spheres*, J. Fluid Mech., 131 (1983), pp. 155-175.

[4] D. DREW, L. CHENG, AND R.T. LAHEY, *The analysis of virtual mass effects in two-phase flow*, Int. J. Multiphase Flow, 5 (1979), pp. 233-242.

[5] D.A. DREW AND R.T. LAHEY, *Application of general constitutive principles to the derivation of multidimensional two-phase flow equations*, Int. J. Multiphase Flow, 5 (1979), pp. 243-264.

[6] D.A. DREW, *Mathematical modeling of two-phase flow*, Ann. Rev. Fluid Mech., 15 (1983), pp. 261-291.

[7] F. DOBRAN, *A two-phase fluid model based on the linearized constitutive equations*, in Advances in Two-Phase Flow and Heat Transfer, vol. 1, S. Kakac and M. Ishii eds., Martinus Nijhoff Publishers, Boston, 1983, pp. 41-59.

[8] G.B. WALLIS, *Inertial coupling in two-phase flow: macroscopic properties of suspensions in an inviscid fluid*, Thayer School of Engineering, Dartmouth College, October 1988.

[9] A.V. JONES AND A. PROSPERETTI, *On the suitability of first-order differential models for two-phase flow prediction*, Int. J. Multiphase Flow, 11 (1985), pp. 133-148.

[10] A. PROSPERETTI AND A.V. JONES, *The linear stability of general two-phase flow models – II*, Int. J. of Multiphase Flow, 13 (1987), pp. 161-171.

[11] T.B. ANDERSON AND R. JACKSON, *Fluid mechanical description of fluidized beds*, I&EC Fundamentals, 7 (1968), pp. 12-21.

[12] W.C. THACKER AND J.W. LAVELLE, *Stability of settling of suspended sediments*, Phys. Fluids, 21 (1978), pp. 291-292.

[13] C.D. HILL AND A. BEDFORD, *Stability of the equations for particulate sediment*, Phys. Fluids, 22 (1979), pp. 1252-1254.

[14] D.A. DREW AND R.T. WOOD, *Overview and taxonomy of models and methods*, presented at the International Workshop on Two-Phase Flow Fundamentals, National Bureau of Standards, Gaithersburg, Maryland, Sept. 22-27,1985.

[15] C. PAUCHON AND S. BANERJEE, *Interphase momentum interaction effects in the averaged multifield model: Part I*, Int. J. Multiphase Flow, 12 (1986), pp. 559-573.

[16] G.B. WALLIS, *On Geurst's equations for inertial coupling in two-phase flow*, this volume.

[17] N. ZUBER, *On the dispersed two-phase flow in the laminar flow regime*, Chem. Eng. Sci., 19 (1964), pp. 897-917.

[18] YU.G. MOKEYEV, *Effect of particle concentration on their drag and induced mass*, Fluid Mech. - Sov. Res., 6 (1977), pp. 161-168.

[19] L. VAN WIJNGAARDEN, *Hydrodynamic interaction between gas bubbles in liquid*, J. Fluid Mech., 77 (1976), pp. 27-44.

[20] R.M. GARIPOV, *Closed system of equations for the motion of a liquid with gas bubbles*, J. of App. Mech. and Tech. Phys, 14 (1973), pp. 737-756. Translated from Z. Prikladnoi Mekhaniki i Teckhnicheskoi Fiziki, 6 (1973) p. 3-24.

[21] R.I. NIGMATULIN, *Spatial averaging in the mechanics of heterogeneous and dispersed systems*, Int. J. of Multiphase Flow, 5 (1979), pp. 353-385.

[22] J.A. GEURST, *Variational principles and two-fluid hydrodynamics of bubbly liquid/gas mixtures*, Physica, 135A (1986), pp. 455-486.

[23] M. ISHII AND N. ZUBER, *Drag coefficient and relative velocity in bubbly, droplet or particulate flows*, AIChE Journal, 25 (1979), pp. 843-855.

MATHEMATICAL ISSUES IN THE CONTINUUM FORMULATION OF SLOW GRANULAR FLOW

DAVID G. SCHAEFFER*

(Dedicated to Roy Jackson, who introduced me to granular flow)

1. Introduction. This lecture is a survey of certain mathematical problems arising in the description of the flow of granular materials such as sand, coal, or various raw materials used in industry. The rather long title is intended to convey what is not covered. We refer to slow flow in contrast to the rapid shearing flow considered by many authors (cf. the article by Jenkins in these Proceedings). In the case of rapid shearing it is possible to derive constitutive relations from first principles; the constitutive relations used here are purely phenomenological. We refer to a continuum formulation in contrast to particle simulations (cf. the article by Walton in these Proceedings). Although two phase flow was a major theme at the Workshop, we shall not consider the interaction of the granular medium with the interstitial fluid. We have retained the informal style of the lecture in this article.

2. Some information from experiments. In this section we call the reader's attention to a few of the interesting properties of granular flows. It is fokelore that the flow from a hopper is pulsating. In experiments reported in these Proceedings, Baxter and Behringer have measured the time varying stresses on a laboratory-sized hopper, taken Fourier transforms, and analyzed the resulting power spectra.

Besides temporal dependence, granular flow can exhibit interesting spatial dependence. In the same article, Baxter and Behringer report of X-ray measurements which show density waves that propagate in a discharging hopper; usually they propagate upwards, but in a hopper with steep walls they propagate downward.

Discontinuities often arise in granular flow. For example they occur at the boundary between stagnant and flowing regions in a hopper with a large opening angle (cf. Figure 1), and they occur in the shear bands which often form in constitutive tests (cf. the article by Vardoulakis in these Proceedings).

3. Derivation of simplest model. For simplicity we shall work in two dimensions. See Prager [5] for a lucid general discussion of the underlying continuum mechanics. The description of the flow requires knowledge of the density ρ, the velocity $\vec{v}$, and the stress tensor T_{ij}. These variables are subject to conservation of mass

$$d_t\rho + \rho\partial_i v_i = 0 \tag{1}$$

conservation of momentum

$$\rho\, d_t v_i + \partial_j T_{ij} = F_i, \tag{2}$$

*Department of Mathematics Duke University, Durham, North Carolina 27706

and appropriate constitutive relations that we shall derive momentarily. Here $d_t = \partial_t + v_j\partial_j$ is the convective derivative, F_i is the body force, and we employ the summation convection.

The constitutive law relates the stress tensor T_{ij} to the strain rate tensor

$$V_{ij} = -\frac{1}{2}\,(\partial_j v_i + \partial_i v_j). \tag{3}$$

(Because of the minus sign in (3), contractive strains correspond to positive eigenvalues of V_{ij}. Likewise we take compressive stresses as positive.) For motivation, let us first recall the constitutive relation of an incompressible Newtonian fluid,

$$T_{ij} = p\delta_{ij} + 2\mu V_{ij} \tag{4}$$

where $p = \frac{1}{2}\, trT$ is the pressure and μ, a physical constant, is the viscosity. It is convenient to rewrite (4) in the form

$$V_{ij} = \frac{1}{2\mu}\,\operatorname{dev} T_{ij}, \tag{5}$$

where the deviator of a 2×2 matrix is defined by

$$\operatorname{dev} A = A - \frac{1}{2}(trA)I.$$

In particular, in an ideal fluid (i.e., $\mu = 0$), a hydrostatic pressure is the only allowable stress. Note that it follows from (5) that

$$\operatorname{div} v = -V_{ii} = 0.$$

Because the fluid is incompressible, the pressure is not specified by the constitutive law. Rather it is obtained by solving the Navier-Stokes equations, which result from substitution of (4) into (1) and (2):

$$\begin{gathered}(a)\ \rho\ d_t v_i = -\partial_i p + \mu\partial_{jj} v_i \\ (b)\ \partial_i v_i = 0.\end{gathered} \tag{6}$$

For a granular material, we need three scalar constitutive laws to supplement (1), (2). The first of these is motivated by the idea that, unlike a fluid, a granular material can support some shearing stress at rest, but only up to a certain threshold. In symbols, this is expressed by Coulomb's yield condition

$$|\operatorname{dev} T| \le k\frac{trT}{2} \tag{7}$$

where k is a constant and $|\cdot|$ is the Euclidean norm

$$|A| = \left\{\sum |a_{ij}|^2\right\}^{1/2}.$$

If T has eigenvalues σ_1 and σ_2, then (7) is equivalent to

$$\sqrt{2}\left(\frac{\sigma_1-\sigma_2}{2}\right) \leq k\,\frac{\sigma_1+\sigma_2}{2}\,.$$

In order to guarantee that both principal stress are compressive (i.e., $\sigma_i > 0$), we shall require that

$$k < \sqrt{2}\,.$$

We shall suppose that the material is rigid-plastic, which means that a material element behaves rigidly if the inequality (7) is strict and can deform only if (7) is satisfied by virtue of equality. As indicated in Figure 2, $|\operatorname{dev} T| = k\; trT/2$ defines a cone in stress space whose axis is the hydrostatic axis. The constitutive assumption excludes stresses outside this cone, just as in an ideal fluid stresses off the hydrostatic axis are excluded.

If material is deforming, then the yield condition provides the constitutive relation

$$|\operatorname{dev} T| = k\,\frac{trT}{2}\,. \tag{8}$$

However, if material is behaving rigidly, then in fact the stress equations are undetermined.

In words, the remaining two constitutive laws are (i) incompressibility and (ii) the eigenvectors of the stress tensor T and the strain rate tensor V are coaxial. The latter condition is motivated by Figure 3 — for example, under a large vertical load and relatively small horizontal load, one expects the material to expand in the horizontal direction as indicated. These two conditions can be contained in the single tensor equation

$$V = q\operatorname{dev} T \qquad (q > 0) \tag{9}$$

for some q, a function of position and time. Although (9) is formally similar to (5), note that q in (9) depends on experimental circumstances while μ in (5) is a physical constant.

We may combine (8) and (9) to express T as a function of V. Now

$$T = \sigma I + \operatorname{dev} T = \sigma I + |\operatorname{dev} T|\,\frac{\operatorname{dev} T}{|\operatorname{dev} T} \tag{10}$$

where $\sigma = trT/2$ is the mean stress. Determining $|\operatorname{dev} T|$ from (8) and deducing

$$\frac{\operatorname{dev} T}{|\operatorname{dev} T|} = \frac{V}{|V|}$$

from (9), we conclude that

$$T = \sigma\left\{I + k\,\frac{\operatorname{dev} V}{|\operatorname{dev} V|}\right\}. \tag{11}$$

Substituting (11) into conservation of momentum (2), we obtain a system describing granular flow and loosely analogous to the Navier-Stokes equations:

$$(12)\qquad \begin{aligned} &(\mathrm{a})\ \rho d_t v_i + \partial_j \left[\sigma \delta_{ij} + k\sigma \frac{V_{ij}}{|V|}\right] &= 0 \\ &(\mathrm{b}) & \operatorname{div} v = 0. \end{aligned}$$

We note two points about (12). first, even though one constitutive assumptions included no viscosity, nonetheless (12a) contains second order space derivatives of the velocity. Second, the expression containing second order derivatives of velocity is homogeneous *of degree zero*. I.e., dissipation is not increased by doubling the velocities. This property, known as rate independence, is characteristic of sliding friction, which is the dominant source of the internal stresses in a slowly deforming granular material.

4. Ill-posedness of this model. Consider equations (12) on all $\mathbf{R}^2$; i.e., on a domain without boundaries. To study ill-posedness, we linearize (12) about a state in which T and V are constant and look for solutions with exponential dependence

$$(13)\qquad e^{i(\xi,x)+\lambda(\xi)t}$$

where $\xi \in \mathbf{R}^2$ is the vector wave number, $(\cdot,\cdot)$ is the inner product, and $\lambda(\xi) \in \mathbf{C}$ is the growth rate to be determined. For simplicity let us approximate the convective derivative d_t by the partial derivative ∂_t. This simplification is not crucial.

We claim that the growth rate $\lambda(\xi)$ is real and homogeneous of degree two, and it is positive or negative in sectors as sketched in Figure 3. Thus $\lambda(\xi)$ is unbounded from above, so (12) is linearly ill-posed. The forward-backwards heat equation

$$u_t = u_{xx} - u_{yy}$$

is a scalar equation with analogous behavior.

Before proving the claim, we observe that only linearized ill-posedness is being studied here, and linearized ill-posedness does not guarantee the same property for the nonlinear equations. Also it is noteworthy (see [8]) that the ill-posedness is less severe in three dimensions.

For guidance in proving the claim, let us recall the analogous calculation for the Stokes equations. A solution of the linearized equations has the exponential dependence (13) iff λ is a generalized eigenvalue of the problem (in block notation)

$$(14)\qquad \begin{pmatrix} -\frac{\mu}{\rho}\,|\xi|^2 I & \xi \\ \xi^T & 0 \end{pmatrix} \begin{pmatrix} v \\ p \end{pmatrix} = \lambda \begin{pmatrix} v \\ 0 \end{pmatrix}.$$

To determine λ, first note from the second component of (14) that $(\xi, v) = 0$. Taking the scalar product of (14) with ξ, we may express p in terms of v (not involving λ). Finally we obtain the eigenvalue

$$\lambda = -\frac{\mu}{\rho}\ |\xi|^2$$

on the one dimensional space $\{\xi\}^\perp \subset \mathbf{R}^2$. In particular, the Stokes equations are linearly well-posed.

In the granular case, on performing the indicated differentiations in (12) and taking the linearizations, we obtain the generalized eigenvalue problem

$$\begin{pmatrix} \frac{k\sigma}{\rho|V|}\left[-\frac{|\xi|^2}{2}\, I + (A\xi)(A\xi)^T\right] & (I+kA)\xi \\ \xi^T & 0 \end{pmatrix} \begin{pmatrix} v \\ \sigma \end{pmatrix} = \lambda \begin{pmatrix} v \\ 0 \end{pmatrix},$$

where $A = V/|V|$. Note that $trA = 0, trA^2 = 1$; thus in coordinates such that T and V are diagonal,

$$A = \begin{pmatrix} \frac{1}{\sqrt{2}} & 0 \\ 0 & -\frac{1}{\sqrt{2}} \end{pmatrix}. \tag{15}$$

Computing λ as before, we find

$$\lambda = -\frac{k\sigma}{\rho|V|}\left\{\frac{(A\xi,\xi)^2 - \frac{1}{2}\,k|\xi|^2(A\xi,\xi)}{|\xi|^2 + k(A\xi,\xi)}\right\}.$$

Since we have required that $k < \sqrt{2}$, the quadratic form in the denominator is positive definite. However the quartic form in the numerator is indefinite — in particular, it contains the factor $(A\xi,\xi)$ which, by (15), is an indefinite quadratic form. This proves the claim.

(Remark: The lines in Figure 3 separating regions where λ is positive from regions where λ is negative are characteristics of the steady state equations, which are hyperbolic.)

What went wrong in this simple model? In the Navier-Stokes equations, the pressure force associated to the constraint $\operatorname{div} v = 0$ can do no work since the gradient of a scalar is orthogonal to a divergence-free vector field. By contrast, the pressure force in (12) can do work, and for plane waves in certain directions, it does so. The work done by this pressure is drawn from the internal energy of the continuum, but the internal energy of an impressible medium is indeterminate, being the product of an infinite bulk modulus time a vanishing change in volume. Thus arbitrarily much energy is available to feed the growth found above.

5. Alternative models. Some ill-posedness seems to be appropriate for equations describing granular flow since discontinuities develop in finite time. However, the ill-posedness in (12) is too severe in that it does not allow for an initial period of homogeneous deformation in constitutive tests. In this section we briefly discuss some alternative constitutive relations.

(a) SPENCER'S MODEL:

The coaxiality condition has always seemed to us to be the most doubtful equation in the constitutive assumption. Spencer [10] proposed an elegant relation based on steady-state considerations which retains isotropy and leads to steady state equations with double characteristics. Thus the wedges in Figure 3 collapse to zero width and λ has constant sign. Unfortunately λ is everywhere *positive*, so this theory is unsuitable for dynamics.

(b) CRITICAL STATE SOIL MECHANICS:

By *normality* one means a flow rule of the form

$$Vij = q\frac{\partial \Phi}{\partial T_{ij}} \tag{16}$$

where $\Phi(T) = 0$ is the yield function and $q > 0$; in words, (16) states that the strain rate tensor is orthogonal in stress space to the yield surface. In a theory of this type, the pressure force cannot do work. Equation (16), taken with respect to the Coulomb yield surface (8), predicts unreasonably large dilations. Critical State Soil Mechanics, which is discussed in the lecture by Collins in these Proceedings, introduces a family of pressure dependent yield surfaces in a way that the flow rule (16) becomes more physical. Although this mathematically elegant theory is accurate for some soils, especially clays, it seems not to be accurate for dry sand. Thus this theory is a promising area for further mathematical analysis, but it cannot provide a full explanation of the data with sand. (Ill-posedness in critical state Soil Mechanics is studied in [4].)

(c) ELASTICITY:

Since the elastic internal energy is the source of the ill-posedness in the model we studied, it is natural to include elastic effects in an elastic-plastic, as opposed to rigid-plastic model. This modifies the constitutive law (9) by the inclusion of time derivatives of the stress on the right. Consequently it is no longer possible to solve for T as a function of V; rather one must leave the equations as a first order system for both v and T. It turns out (see [2, 7]) that this effect makes little difference regarding well-posedness. This is perhaps to be expected since the elastic moduli of a typical granular medium are much greater than the plastic yield stress — the former are associated with deformation of the solid grains while the latter are associated with the grains sliding over one another.

(d) SHEAR-STRAIN HARDENING:

A less obvious effect promoting well-posedness is shear-strain hardening, i.e., as illustrated in Figure 4, the stress required to plastically deform a sample increases with the accumulated deformation, at least initially. In symbols this requires a variable yield surface.

$$\Phi(T, \gamma) = 0 \tag{17}$$

where γ is the accumulated shear-strain defined by

$$d_t\gamma = |\operatorname{dev} V|. \tag{18}$$

It was shown by Mandel [2] (see also [7]) that the combination of *both* shear-strain hardening and elasticity can lead to a well-posed model. Loosely speaking, the equations are well posed if the product $|C| \left|\frac{\partial\Phi}{\partial\gamma}\right|$ exceeds a constant times the square of the angle of deviation from normality, where the compliance matrix C is the inverse of the elastic moduli and the derivative $\partial\Phi/\partial\gamma$ is the hardening modulus.

(e) HIGHER ORDER CONTINUUM:

By a higher order continuum we mean a theory which includes another continuum variable representing some aspect of the microstructure of the granular medium. Two such theories seem relevant for granular flow: (i) The Cosserat Theory (cf. [3]) which allows the rotation of individual grains to differ from $\frac{1}{2}$ curl v and (ii) a theory (cf. [1]) which allows for fluctuations of particle velocities about a mean. Preliminary research suggests that both of these effects regularize the equations, in a fashion analogous to the way viscosity regularizes the equations of compressible gas dynamics. It is not clear under what circumstances these effects need to be included in the modeling.

6. Issues for further work. The mathematical study of granular flow is just beginning, and far more remains to be done than what has been completed so far. In this section, we mention a few such areas beyond the obvious issues of combining the various effects mentioned in the previous section.

(a) STABILITY VS. ILL-POSEDNESS:

We call a system linearly well-posed if

$$\sup_{\xi} Re\lambda(\xi) < \infty$$

and linearly stable if

$$\sup_{\xi} Re\lambda(\xi) \leq 0.$$

It is natural to expect that a system will become unstable before it becomes ill-posed. As shown in [7, 9], this expectation is borne out in at least two cases. This issue needs to be explored more fully, especially as regards implications for experiments.

(b) BOUNDARY CONDITIONS

All the above analysis was for problems on an infinite space. To model specific constitutive tests one needs to study a problem on a finite domain with appropriate boundary conditions.

(c) INDUCED ANISOTROPY:

The hardening assumed in (17, 18) is isotropic. In reality, a shearing granular medium becomes stronger in the direction in which it is being sheared and weaker in transverse directions. In other words, the hardened yield surface has the anisotropic shape illustrated in Figure 5. The stability of models including this effect need to be investigated.

(d) Rigid regions:

As mentioned above, the equations for stresses in a rigid-plastic medium are undetermined if the medium is rigid. As illustrated in Figure 1, this may occur in a hopper with gently sloped walls. The equations become determined if elasticity is included.

A simple analogy may help illustrate this. Consider the rigid bar supported at three equally spaced points, as illustrated in Figure 6. It is statically indeterminant how the weight of the bar will be distributed between the three supports — to calculate this one must calculate the elastic deflections of the beam. Moreover, the result will be greatly changed by imperfections such as a small curvature in the undeformed state of the bar. In an engineering application it would seem wise to allow for the full range of statically admissible solutions. Arguing by analogy, it would seem worthwhile to study the full range of stresses consistent with the rigid-plastic theory.

(e) Fluid in the interstices:

Most soils contain water in the pores, and the interaction between the fluid and solid phases affects the strength of the soil. In industrial applications of granular flow air is usually the interstitial fluid, but again the fluid-solid interaction can affect the solids flow, especially for fine powders or in pressurized vessels. The interaction between the two phases is a challenging area requiring further study. (See the lecture by Vardoulakis in these Proceedings for preliminary work in this direction.)

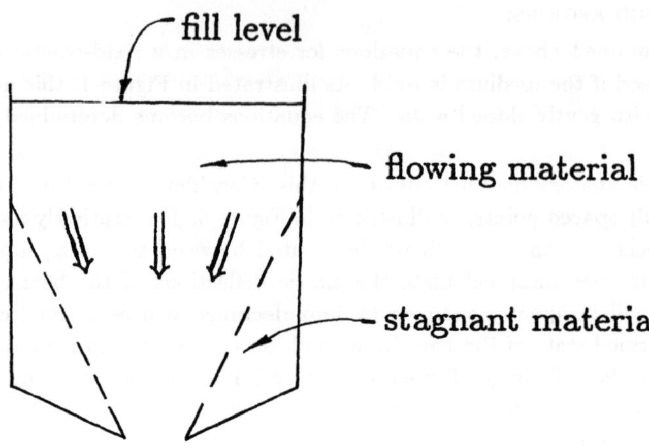

FIGURE 1: A silo with stagnant regions

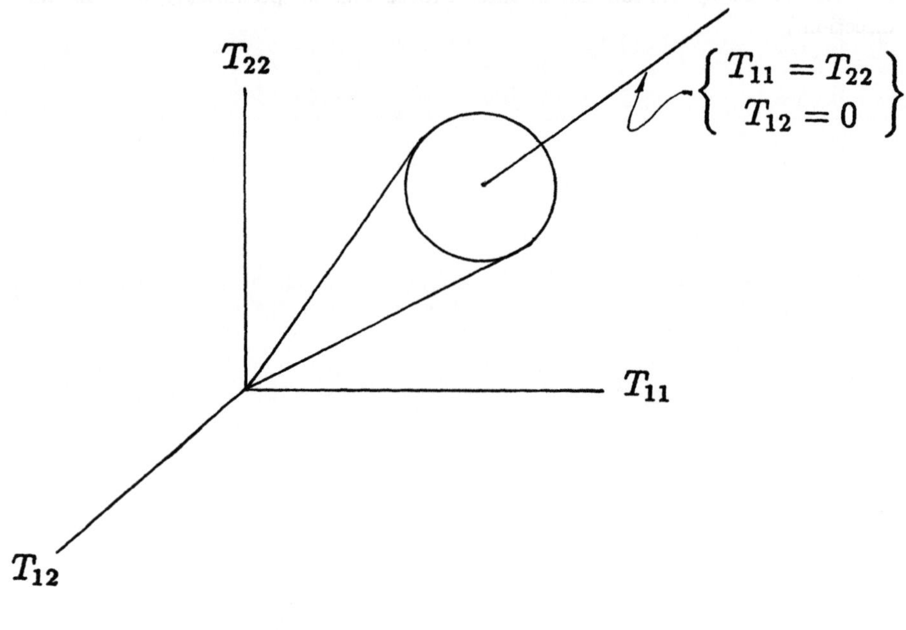

FIGURE 2: The yield cone in stress space

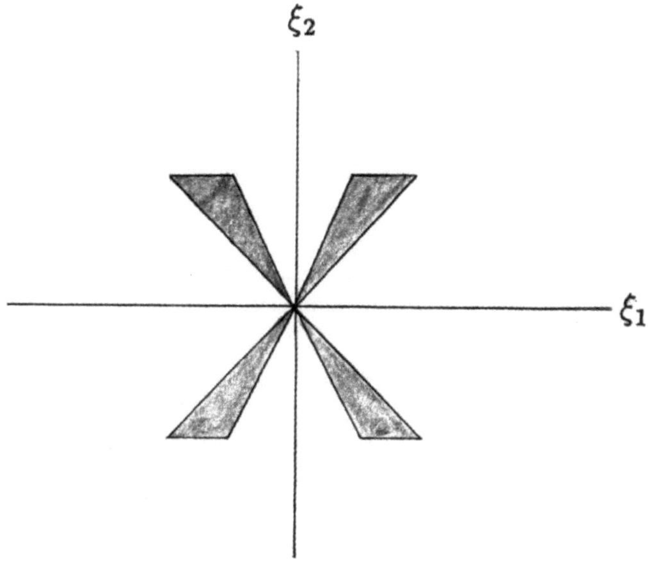

FIGURE 3: Wedges in which the growth rate $\lambda(\xi)$ is unbounded. (Note: The axes are chosen so that the stress tensor is diagonal.)

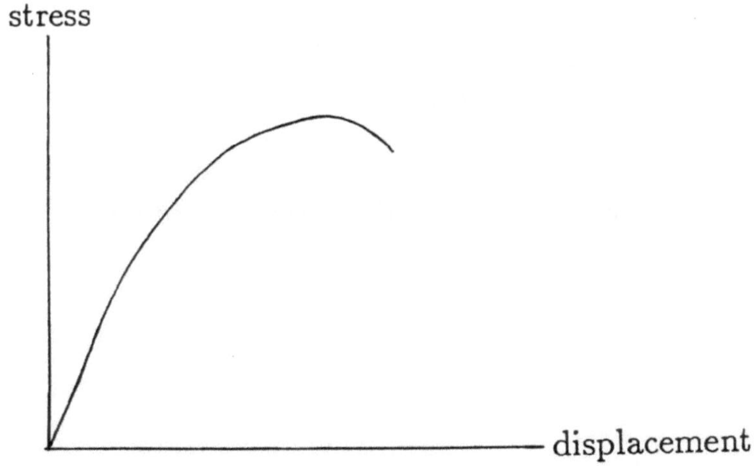

FIGURE 4: Stress vs. displacement in plastic yield.

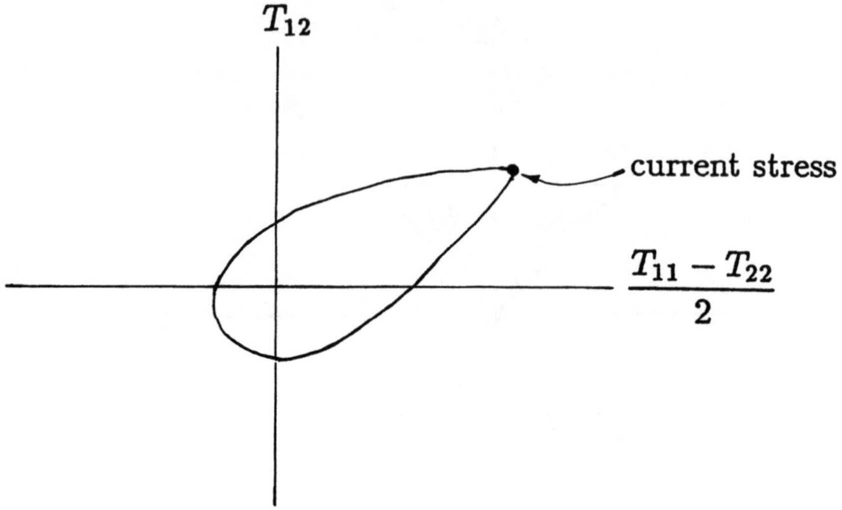

FIGURE 5: A possible anisotropic yield surface. (Note: The hydrostatic axis is suppressed in this figure. The graph above is the intersection of a plane $T_{11} + T_{22} =$ const. with the yield surface.)

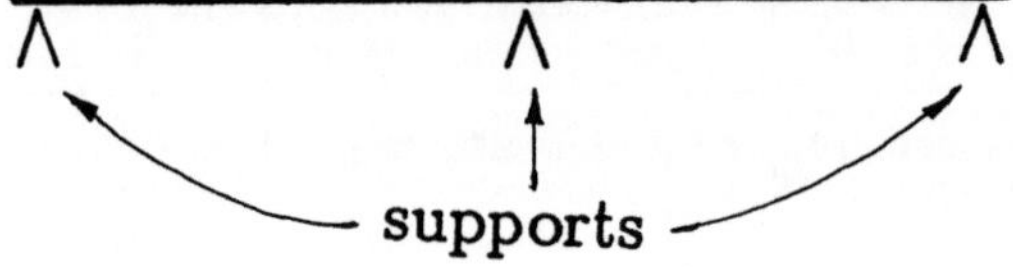

FIGURE 6: A simple statically indeterminant problem

REFERENCES

[1] P.C. Johnson and R. Jackson, *Frictional-collisional constitutive relations for granular materials, with applications to plane shearing*, JFM 176 (1987), pp. 07–93.

[2] J. Mandel, *Conditions de Stabilié et Postulat de Drucker, in Rheology and Soil Mechanics*, G. Kraotchenko and P. Sirieys (eds.), IUTAM Symposium at Grenoble, April (1964), pp. 58–68.

[3] H.-B. Mühlhaus and I. Vardoulakis, *The thickness of shear-bands in granular materials*, Géotechnique 37 (1987), pp. 271–283.

[4] E.B. Pitman and D. Schaeffer, *Stability of time dependent granular flow in two dimensions*, Comm. Pure Appl. Math 40 (1987), pp. 421–447.

[5] W. Praeger, *Introduction to the Mechanics of Continua*, Giun, Watham MA (1961).

[6] D. Schaeffer, *Instability in the evolution equations describing incompressible granular flow*, J. Diff Eq 66 (1987), pp. 19–50.

[7] D. Schaeffer, *Instability and ill-posedness in the deformation of granular materials*, Int'l J. Num. Anal. Methods Geomech., (to appear).

[8] D. Schaeffer and E. B.P. Pitman, *Ill-posedness in three dimensional plastic flow*, Comm. Pure Appl. Math. 41 (1988), pp. 879–890.

[9] D. Schaeffer, M. Shearer, E.B. Pitman, *Instability in critical state theories of granular flow*, SIAM J. Appl. Math, (to appear).

[10] A.J.M. Spencer, *A theory of the kinematics of ideal soils under plane strain conditions*, J. Mech Phys. Solids 12 (1964), pp. 337–351.

ONE-DIMENSIONAL, PARTICLE BED MODELS OF FLUIDIZED SUSPENSIONS*

P. SINGH† AND D.D. JOSEPH†

Abstract. One-dimensional unsteady models of a fluidized suspension based on modeling the forces that the fluid exerts on the particles are considered. Four different theories are discussed. The first, by Foscolo and Gibilaro [1984, 1987], gives a criterion for the loss of stability of uniform fluidization. A second theory by Joseph [1990] which appears to carry the Foscolo-Gibilaro theory to a logical conclusion with the addition of a term proportional to the particle velocity gradient, leads always to instability. A third theory by G.K. Batchelor [1988] is formally similar to the one by Foscolo-Gibilaro, but is more generally derived. A fourth theory which takes into account the finite size of particles and can be used in any of the other three theories is derived here. We show that the finite size of particles is a regularizer of the short wave instability of uniform fluidization which occurs when the particle phase pressure is neglected. We introduce the problem of losing range. If the fluids and solids fractions are both intitially in the interval (0, 1), will they stay on that interval as they evolve? An answer is given.

1. Fluidized Beds. A particle is fluidized when it is lifted against gravity by the drag of upward moving fluid. The particle is in equilibrium under weight and drag. An assemblage of particles in a container which is not fluidized rests on the bottom of the container. Below this speed, fluid passing up through such a bed will see the bed as a porous media. There is a critical speed above which the particles are fluidized. The bed expands to maintain a balance between drag and weight when the flow rate is increased. A statistically homogeneous fluidized bed with constant flow throughput is called a state of uniform fluidization. Such states are notoriously difficult to achieve. It appears to be true that gas fluidized beds of light particles can be stable above minimum fluidization. When the flow throughput is large the gas collects into large gas bubbles which rise through the bed. This is a failure of fluidization since the individual particles are basically insufficiently fluidized to promote efficient heat and mass transfer. The transition to bubbling is said to be a transition from particulate to aggregate fluidization, the particles aggregate, with gas in clear regions. It is not clear that uniform fluidization is the same as particulate fluidization; for example, waves may appear destroying uniformity without marked aggregation of particles. The concept of stability itself is not clear since stability is defined only in a statistical sense. The particles are probably always shaking about.

Stability analysis for bubbling beds is a matter of great interest for the technology of beds used in catalytic cracking and coal combustion. It also plays a certain role in the theory of multiphase flow as a test problem.

---

*This research was supported under grants from the US Army Research Office, Mathematics, the Department of Energy and the National Science Foundation.

†Department of Aerospace Engineering and Mechanics, University of Minnesota, 107 Akerman Hall, Minneapolis, MN 55455

2. Two-fluid Equations. We can form continuum equations for two-fluids, even when one of the two constituents is solid, by ensemble averaging (Drew, 1983). Joseph and Lundgren [1990] have derived the following set of ensemble averaged equations for incompressible fluid-particle suspensions.

$$\frac{\partial \varepsilon}{\partial t} + \operatorname{div} \varepsilon \mathbf{u}_f = 0 \ , \tag{1}$$

$$\frac{\partial \phi}{\partial t} + \operatorname{div} \phi \mathbf{u}_p = 0 \ , \tag{2}$$

$$\begin{aligned} \rho_f \varepsilon \left(\frac{\partial \mathbf{u}_f}{\partial t} + \mathbf{u}_f . \nabla \mathbf{u}_f \right) &+ \rho_f \operatorname{div} \langle H(\mathbf{V} - \mathbf{u}_f)(\mathbf{V} - \mathbf{u}_f) \rangle \\ &= -\nabla(p_f \varepsilon) + \mu \nabla^2 \mathbf{u}_c - \langle \delta_\Sigma(\mathbf{x}) \mathbf{t} \rangle + \rho_f \varepsilon \mathbf{b}_f \ , \end{aligned} \tag{3}$$

$$\begin{aligned} \rho_p \phi \left(\frac{\partial \mathbf{u}_p}{\partial t} + \mathbf{u}_p . \nabla \mathbf{u}_p \right) &+ \rho_p \operatorname{div} \langle (1-H)(\mathbf{V} - \mathbf{u}_p)(\mathbf{V} - \mathbf{u}_p) \rangle \\ &= -\nabla(p_p \phi) + \langle \delta_\Sigma(\mathbf{x}) \mathbf{t} \rangle + \rho_p \phi \mathbf{b}_p \ , \end{aligned} \tag{4}$$

where

$H(\mathbf{x})$ is an indicator function, 0 if $\mathbf{x}$ is in the fluid, 1 otherwise
$\langle . \rangle (\mathbf{x}, t)$ ensemble average
$\varepsilon = \langle H \rangle$ fluid fraction
$\phi = \langle 1 - H \rangle$ solid fraction
$\mathbf{u}_f = \langle H\mathbf{V} \rangle$ average fluid velocity, where $\mathbf{V}$ is the true velocity
$\mathbf{u}_p = \langle (1-H)\mathbf{V} \rangle$ average particles velocity
$(1-\phi)p_f = \langle Hp \rangle$ fluid phase pressure, where p is the mean normal stress
$\phi p_p = \langle (1-H)p \rangle$ particle phase pressure
$\mathbf{u}_c = \varepsilon \mathbf{u}_f + \phi \mathbf{u}_p$
τ is the extra stress $\mathbf{T} = -p\mathbf{l} + \tau$
$\mathbf{b}_f$ and $\mathbf{b}_p$ ensemble averaged body forces in the fluid and solid
δ_Σ is a Dirac delta function across the solid-fluid interface
$\mathbf{t} = \mathbf{n}.\mathbf{T}$ is the traction vector on the solid-fluid interface.

If we add equations (1) and (2) we get

$$\operatorname{div} \mathbf{u}_c = 0 \ . \tag{5}$$

The boundary conditions between the fluid and the particle involves the traction vector term in (3) and (4) and it is probably best not to combine the two equations.

The existence of two fluid equations even when one of the fluids is solid is perfectly justified by ensemble averaging. These equations, like other two fluid models, are not closed and methods of closure, or constitutive models for the interaction terms, are required to put the equations into a form suitable for applications. Moreover, since averaging over repeated identical trials is not a realizable proposition, the ensemble average variables are conceptually abstract and their relation to more physically intuitive variables, like the ones which arise from spatial averaging, is uncertain.

Equations (1) through (5) are appropriate for fluidized suspensions with $\mathbf{b}_f = \mathbf{b}_p = \mathbf{g}$, gravity. Particle bed models decouple the fluids and solids equations and work with the solids equations alone.

3. On Losing Range. One of us (DDJ) was worried for a week in June 1989 about the possibility of losing range. The range $0 < \varepsilon(\mathbf{x}, t) < 1$ must be preserved by dynamics; if $\varepsilon(\mathbf{x}, 0)$ is between 0 and 1 is it possible for ε to go negative or grow larger than one? This should not happen; moreover, the protection of the range should not depend on $\mathbf{u}_f$ and $\mathbf{u}_p$ because these fields can be changed at will by changing the constitutive equations or initial conditions. The problem is this: given equations (1) and (2) and sufficiently smooth field $\mathbf{u}_f$ and $\mathbf{u}_p$ and $0 < \varepsilon(\mathbf{x}, 0) < 1$, what are the conditions such that $0 < \varepsilon(\mathbf{x}, t) < 1$ for all t. This problem was given to Sir James Lighthill, who gave the following solution (in one space dimension). We write

$$\frac{\partial \varepsilon}{\partial t} + u_f \frac{\partial \varepsilon}{\partial x} + \varepsilon \frac{\partial u_f}{\partial x} = 0 \tag{6}$$

and deduce that along a curve

$$\frac{dx}{dt} = u_f \tag{7}$$

we have

$$\frac{d\varepsilon}{dt} = -\varepsilon \frac{\partial u_f}{\partial x} \, . \tag{8}$$

Therefore, if $\varepsilon = \varepsilon_0$ on this curve at $t = 0$, then we have

$$\varepsilon = \varepsilon_0 \exp\left(- \int_0^t \frac{\partial u_f}{\partial x} \, dt \right) \tag{9}$$

which is always positive since $\varepsilon_0 > 0$. Similarly we have

$$\frac{d(1-\varepsilon)}{dt} = -(1-\varepsilon) \frac{\partial u_p}{\partial x} \, . \tag{10}$$

Then, if $\varepsilon = \varepsilon_0$ on this curve at $t = 0$, we have

$$1 - \varepsilon = (1 - \varepsilon_0) \exp\left(- \int_0^t \frac{\partial u_p}{\partial x} \, dt \right) \tag{11}$$

which is always positive.

This proof works for the three dimensional case with $\frac{\partial u}{\partial x}$ under the integral replaced by div $\mathbf{u}$. It also works for compressible constituents, not discussed here.

It is clear that we can protect the range if and only if both (1) and (2) are satisfied. In particle bed models only (2) is satisfied, so we are in danger of losing range.

4. Particle Bed Models. If we knew the force on each and every particle we could in principle track their motion. The ultimate in particle bed calculations would be a molecular dynamic simulation. For this to work we would need to know the force that the fluid exerts on the particles. There is no perfect way to do this without doing the fluid dynamics. In fact, exact numerical solutions correlating certain small motions of particles with forces generated by the flow of a Navier-Stokes fluid is a viable proposition (for example, see Singh, Caussignac, Fortes, Joseph and Lundgren [1989]) with a great future. Unfortunately it is not possible to know perfectly how the fluid forces will effect a particle without actually doing the fluid mechanics. The hope behind the particle bed models discussed below is that our experience and undertanding of fluid-particle interactions will allow us to guess correctly what form these interactions ought to take, at least in an average sense. It is by no means certain that this hope can be realized.

One-dimensional particle bed models have been given by Foscolo and Gibilaro [1984, 1987] and G.K. Batchelor [1988]. These theories will be reviewed below. They use mass conservation of the solid in the ensemble averaged form (2)

$$\frac{\partial \phi}{\partial t} + \operatorname{div} \phi \mathbf{u}_p = 0 \ .$$

This equation does not acknowledged either structure or size effects of particles on the continuity of flow. Approaches, like the one given below, based on geometry rather than ensemble averaging may be preferred.

5. Mass Balance Equations for Balls of Radius R. We are going to derive a one-dimensional mass balance for spherical particles of uniform radius R. Consider a plane at $z = Z$, perpendicular to gravity, in figure 1. Let us consider the area $A = L^2$ of a square in this plane with $L \gg R$, so many spheres intersect the plane at Z. Let x be the distance from the plane $z = Z$. All spheres whose centers are at $|x| \leq R$ pass through the plane Z. Spheres with $|x| \geq R$ do not touch Z. The area of the hole cut out by the sphere at $|x| \leq R$ is $\pi(R^2 - x^2)$.

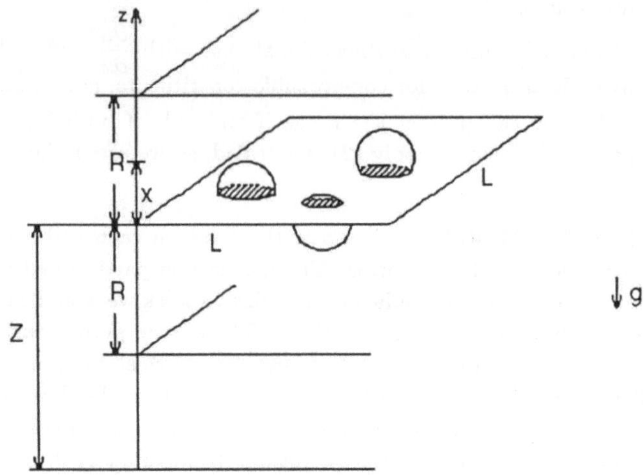

FIGURE 1. Cross-section of fluidized spheres of radius R in the plane at $z = Z$.

Now we define a wafer of influence for the plane at z. Its area is A and thickness is $2R, -R \leq x \leq R$. The total volume of the waver is $2RA$.

The next quantity to be defined is the number density per unit area

$$N(z+x,t) = \frac{\text{number of spheres whose centers are at } z+x}{A}$$

$$= \frac{\text{number of holes in the plane } Z \text{ of area } \pi(R^2 - x^2)}{A} \quad (12)$$

Let V_s be the volume of spheres in the waver of influence and $\phi = \dfrac{V_s}{2RA}$ be the solid fraction. Then

$$dV_s = N(z+x,t)A\{\pi(R^2 - x^2)\}dx$$

is the element of solids volume swept out by number of holes of area $\pi(R^2 - x^2)$ times the volume of one of these holes as x moves through dx. Then

$$V_s = \int_{-R}^{R} N(z+x,t)A\pi(R^2 - x^2)dx \quad (13)$$

and

$$\phi = \frac{1}{2R} \int_{-R}^{R} N(z+x,t)\pi(R^2 - x^2)dx \,. \quad (14)$$

If $N = N_0$ is constant, then

(15) $$\phi_0 = \frac{\left(\frac{4}{3}\pi R^3\right)(N_0 A)}{2RA} = \frac{2}{3}\pi R^2 N_0$$

is the solid fraction for a uniform distribution of spheres.

Now we write out a mass balance. Let $u_p(z+x,t)$ be the velocity of a sphere whose center is at $z+x$. Then

$$A\ N(z+x,t)u_p(z+x,t)\pi(R^2-x^2)$$

is the flux of area through the plane at z of spheres centered at $z+x$ and the increase of concentration at z is balanced by the fluxes of all areas of spheres intersecting z; that is

(16) $$\frac{\partial}{\partial t}\int_{-R}^{R} N(z+x,t)(R^2-x^2)dx + \frac{\partial}{\partial z}\int_{-R}^{R} N(z+x,t)u_p(z+x,t)(R^2-x^2)dx = 0\,.$$

6. The Particle Bed Model of G.K. Batchelor. Batchelor [1988] established the form of the momentum equation for one-dimensional unsteady mean motion of solid particles in a fluidized bed or sedimenting dispersion from physical arguments. He works with area averaged quantities which because of statistical homogeneity can be identified with ensemble averages. He asserts a definite point of view preferring to establish equations carefully, with plausible physical reasoning and a minimum of hypotheses concerning the relation between mean quantities. He avoids the introduction of any parameters that do not have a clear physical meaning and are not calculable or measurable, at least in principle. He obtains the following differential equations for the area averaged mean quantities balancing momentum

(17)

$$mn(1+\theta)\left(\frac{\partial V}{\partial t}+V.\nabla V\right) = -\frac{\partial(mn\langle v^2\rangle)}{\partial x} + n\{F_h(V,\phi)-F_h(U,\phi)\} - \frac{D}{B}\frac{\partial n}{\partial x} + \frac{\partial\left(\phi\rho_f\eta'\frac{\partial V}{\partial x}\right)}{\partial x}$$

where

m = mass of particle,

n local number density, $mn = \rho_p \phi$,

$\theta = \frac{\rho_f}{\rho_p} C$ virtual mass term, where $C = C(\phi)$ depends on ϕ. But here we take C to be a constant for convenience which makes the term dependent on the derivative of C to drop out.

V mean particle velocity. The axis of reference is such that the mean of material volume across a horizontal plane is zero.

When $V \neq 0$; the axis of reference moves with a constant velocity,

v velocity fluctuation with a zero horizontal average, $\langle v \rangle = 0$.

Horizontal averages are assumed to be same as ensemble averages,

$F_h(V, \phi)$ mean force exerted by the fluid on a particle whose mean velocity is V in a homogeneous dispersion of concentration ϕ,

(18) $F_h(U, \phi) = -m\tilde{g}$

where $\tilde{g} = g\frac{\rho_p - \rho_f}{\rho_p}$ is the reduced buoyancy and $V = U(\phi)$ is the mean partic velocity in a uniform bed in which the particles are in equilibrium under weight and drag

B bulk mobility. This is the ratio of the small change of velocity produced by a small change of force,

D local hydrodynamic diffusivity. $\frac{D}{B}$ is a diffusivity coefficient,

$\phi \rho_f \eta'$ a viscosity coefficient.

He considers first the approximate form of the momentum equation when the departure from homogeneity is small and the spatial gradients $\frac{\partial \phi}{\partial x}$ and $\frac{\partial V}{\partial x}$ are small in some sense. Withoug going very deeply into these approximations we list them below: First

$$F_h(V, \phi) - F_h(U, \phi) = \gamma \frac{V - U}{U} F_h(U) = -\gamma m \tilde{g} \frac{V - U}{U} \tag{19}$$

where $F_h(V) = F_h(U) + \frac{\partial F_h}{\partial V}(V - U)$ defines γ, believed to be slowly varying in V and ϕ, and (18) is used to eliminate $F_h(U)$. Second

$$\langle v^2 \rangle = H(\phi) U^2 - \eta''(\phi) \frac{\partial V}{\partial x} - \eta'''(\phi) U \frac{\partial \phi}{\partial x} \tag{20}$$

where following kinetic theory $\eta'''(\ll \eta'' > 0)$ is discarded, $H(0) = 0$ and $H(\phi_0) = 0$ where ϕ_0 is for the close packing because fluctuations must vanish in these two limits. After inserting these approximations, Batchelor obtains

(21)

$$\phi(1+\theta)\left(\frac{\partial V}{\partial t} + V.\nabla V\right) = -\frac{d(\phi H U^2)}{d\phi}\frac{\partial \phi}{\partial x} - \frac{\gamma \tilde{g}}{U}\left(\phi(V - U) + D\frac{\partial \phi}{\partial x}\right) + \frac{\partial \left(\phi \eta \frac{\partial V}{\partial x}\right)}{\partial x}$$

where $\rho_p\eta = \rho_p\ \eta'' + \rho_f\ \eta'$ can be called the particle viscosity and $\rho_p\ \eta''$ is an eddy viscosity.

To compare Batchelor's theory with that of Foscolo and Gibilaro it is convenient to make a Galilean transformation to a laboratory fixed frame from the zero material flux axis. Thus

$$x = x_0 + U_0 t, \quad V = U_0 + u_p$$

where $U_0 = U(\phi_0)$ is independent of x and t and $U_0 = u_c$ is the composite or the fluidizing velocity. Then we have equation (2) and

$$\phi(1+\theta)\left(\frac{\partial u_p}{\partial t} + u_p.\nabla u_p\right) = -Q\frac{\partial \phi}{\partial x} - \frac{\gamma\tilde{g}}{U}\ \phi(U_0 - U - u_p) + \frac{\partial(\left(\phi\eta\frac{\partial u_p}{\partial x}\right)}{\partial x} \tag{22}$$

where

$$Q = \frac{d(\phi H U^2)}{d\phi} + \frac{\gamma\tilde{g}}{U}\ D.$$

In the next section we will present some results which suggest that there ought to be a term proportional to $\dfrac{\partial u_p}{\partial x}$ on the left of (22).

The total coefficient of $-\dfrac{\partial \phi}{\partial x}$ in (22) can be interpreted as a bulk modulus of elasticity of the configuration of particles. This term could also be identified as arising from a particle phase pressure. Batchelor regards the contribution of the part proportional to D to be the more important of the two. He thinks of this as new contribution representing the diffusion of particles against a gradient. We will see later how these gradient terms regulate the short wave instabilities which arise when the derivative terms on the left of (22) are put to zero.

7. The Particle Bed Model of Foscolo-Gibilaro. Foscolo and Gibilaro [1984] start with coupled one-dimensional equations for the particles and fluid phase. The particle phase equations are

$$\frac{\partial \phi}{\partial t} + \frac{\partial \phi u_p}{\partial z} = 0, \tag{23}$$

$$\phi\rho_p\left[\frac{\partial u_p}{\partial t} + u_p\frac{\partial u_p}{\partial z}\right] = -\phi\rho_p g + \mathcal{F} - \frac{\partial p_p}{\partial z}\ . \tag{24}$$

where $\mathcal{F}$ is the interaction force, the force that the fluid exerts on the particle, and p_p is the particle phase pressure. The fluid equations are of the same form except that the subscript p is replaced by f, ϕ is replaced by ε and $\mathcal{F}$ by minus $\mathcal{F}$.

Foscolo and Gibilaro modeled the interaction force $\mathcal{F}$ and the particle phase pressure in a manner that decouples the equations for the fluid and solid phases. This gives rise to a system of equations for the particles only, called the particle bed model.

It is convenient to introduce a dynamic pressure π_p into (24) by writing

$$p_p == P + \pi_p$$
$$\phi p_p g + \frac{\partial P}{\partial z} = 0.$$

Then (24) reduces to

$$\phi\rho_p\left[\frac{\partial u_p}{\partial t}+u_p\frac{\partial u_p}{\partial z}\right]=\mathcal{F}-\frac{\partial \pi_p}{\partial z}\,. \tag{25}$$

To get their equations they first derived an interesting expression $F_d(\varepsilon)$ for the drag force exerted by the fluid on a single particle in a uniform fluidized suspension. This expression relies strongly on the well-known correlation of Richardson and Zaki for fluidized and sedimenting beds of monosized spherical particles

$$u_c = V\varepsilon^n \tag{26}$$

where

$$u_c = u_p\phi + u_f\varepsilon \tag{27}$$

is the composite velocity, the volume flux divided by total area and u_c is independent of z, $\frac{\partial u_c}{\partial z}=0$. Of course $V=u_f$ when $\varepsilon=1$,the steady terminal velocity of a freely falling single sphere in a sea of fluid. The exponent n depends on the Reynolds number $Re=\frac{dV}{\nu}$ where d is the diameter

$$n=\begin{cases}4.65 \text{ for } Re<0.2,\\ 4.4\ Re^{-0.03} \text{ for } 0.2<Re<1,\\ 4.4\ Re^{-0.1} \text{ for } 1<Re<500,\\ 2.4 \text{ for } Re>500.\end{cases} \tag{28}$$

Foscolo–Gibilaro replace 4.65 with 4.8=2(2.4) for reasons to be made clearer later.

There is a huge amount of fluid mechanics buried in the Richardson–Zaki correlation. This is hidden in the drag law for particles falling under gravity in steady flow. Let $F_d(\varepsilon)$ be the drag on a single particle in a freely falling suspension with a fluid fraction ε. When $\varepsilon=1$ we get a drag law for the free fall of a single sphere which is Stokes drag when V is small enough; for larger V the drag is given by

$$F_d(1)=\frac{\rho V^2}{2}\frac{\pi d^2}{4}C_D \tag{29}$$

where C_D is given by an empirical correlation. Foscolo and Gibilaro produce the formula

$$F_d(\varepsilon)=\varepsilon F_d(1) \tag{30}$$

from an argument which says that in a fluidized bed in a steady flow, the total force F on a sphere is the sum

$$F(\varepsilon,Re)=F_d(\varepsilon)-F_p(\varepsilon)$$

where

$$F_p(\varepsilon) = \frac{\pi d^3}{6}(\rho_p - \rho_c)g$$

is the buoyant force using the effective density

$$\rho_c = \varepsilon\rho_p + \phi\rho_p$$

of the composite fluid. Since $\phi = 1 - \varepsilon$,

$$F_p(\varepsilon) = \frac{\pi d^3}{\rho}(\rho_p - \rho_f)g\varepsilon = \varepsilon F_p(1)$$

in steady flow, $F = 0$ and

$$F_d(\varepsilon) = F_p(\varepsilon) = \varepsilon F_p(1) = \varepsilon F_d(1) \ . \tag{31}$$

We never see steady flow in a fluidized bed, the particles always jiggle about; steady is in some statistical sense, whatever that may be. In any interpretation

$$u_p = 0 \quad \text{in steady flow.}$$

Equation (31) is all that is required to get the drag on a single particle in a fluidized suspension in steady flow. The hydrodynamic content is all buried in the drag correlation (29). We may write $F_d(\varepsilon) = \varepsilon F_d(1)$. To see how $F_d(\varepsilon)$ depends on the fluidizing velocity u_c, Foscolo and Gibilaro note that (29) implies that

$$F_d = \varepsilon \begin{cases} 3\pi\mu V & \text{(laminar)} \\ 0.055\pi\rho d^2 V^2 & \text{(turbulent).} \end{cases}$$

They next note that in the Richardson and Zaki correlation (26) and (28), with 4.8 replacing 4.65, implies that

$$F_d = \varepsilon^{-3.8} \begin{cases} 3\pi\mu d u_c & \text{(laminar)} \\ 0.055\pi\rho d^2 u_c^2 & \text{(turbulent).} \end{cases} \tag{32}$$

This is good, we have $F_d(u_f, \varepsilon) = \varepsilon^{-3.8} F_d(u_f)$, independent of V for low and high Reynolds numbers. Now we look for an equivalent expression, valid for all Reynolds numbers in steady flow and

$$F_d(\varepsilon) = F_d(\varepsilon, u_f, V) = \varepsilon^{-3.8} g(u_f, V)$$

which will reduce to (32) at low and high Re. Clearly

$$g(u_c, V) = \varepsilon^{4.8} F_d(1) = \left(\frac{u_c}{V}\right)^{\frac{4.8}{n}} F_d(1).$$

Hence

$$F_d(\varepsilon, u_c, V) = \varepsilon^{-3.8} \left(\frac{u_c}{V}\right)^{\frac{4.8}{n}} F_d(1). \tag{33}$$

This is just another way of writing $F_d(\varepsilon) = \varepsilon F_d(1)$ when 4.65 is replaced with 4.8 which is useful in motivating the constitutive equation (34) below.

Foscolo and Gibilaro assume that in unsteady flow the force on a particle is given by the expression (33) with u_c replaced by the slip velocity

$$u_c - u_p = (1-\varepsilon)u_p + \varepsilon u_f - u_p = \varepsilon(u_p - u_f).$$

Then the unsteady drag force is

$$F_d(\varepsilon, u_c - u_p, V) = \varepsilon^{-3.8} \left(\frac{u_c - u_p}{V}\right)^{\frac{4.8}{n}} F_d(1). \tag{34}$$

In steady flow, $u_p = 0$, and (34) reduces to

$$F_d(\varepsilon) = \varepsilon F_d(1)$$

where balancing drag and buoyancy for a single sphere gives

$$F_d(1) = \frac{\pi d^3}{6}(\rho_p - \rho_f)g.$$

The total force on single particle in a fluidized suspension is given by

$$F = F_d - F_b = \frac{\pi d^3 g}{6}(\rho_p - \rho_f)\left\{\varepsilon - \left[\frac{u_c - u_p}{V}\right]^{\frac{4.8}{n}} \varepsilon^{-3.8}\right\}.$$

The force per unit volume due to all n spheres is

$$\mathcal{F} = NF$$

where

$$N = \frac{\phi}{\pi d^3/6} = \frac{n}{\text{volume}}.$$

Hence, the total force on the particles per unit volume is

$$\mathcal{F} = -\phi(\rho_p - \rho_f)g\left\{\varepsilon - \left[\frac{u_c - u_p}{V}\right]^{\frac{4.8}{n}} \varepsilon^{-3.8}\right\}. \tag{35}$$

In steady flow, u_p and $F = 0$.

To compare their theory, Foscolo and Gibilaro need to model the particle phase pressure and they do so, but their argument is unclear. Their final expression appears to leave out terms that ought to be included. This issue was addressed in a recent paper by Joseph [1990] which is discussed below.

The same force of the fluid on the particles acts at the boundary to keep the particles from dispersing. However, we need to multiply the force on a single particle by the number N_A per unit area

$$N_A = \frac{\phi}{\pi d^2/4} \,.$$

Hence, the dynamic pressure is given by

$$\pi_p = N_A F = \frac{N_A}{N} \mathcal{F} = \frac{2}{3} d\mathcal{F} \,. \tag{36}$$

The idea of making a constitutive equation for the pressure is more allied to gas dynamics where the pressure is a state variable than to incompressible fluid mechanics. In discussing forces which fluids exert on particles G.K. Batchelor noted that in his list (2.3) of forces there is a "...mean force exerted on particles in this volume by the particles outside the volume." Further he notes that the nature of these two forces

> "...may be explained by reference to a hypothetical case in which the particles are electrically charged and exert repulsive electrostatic forces on each other. The range of action of these electrostatic forces is small by comparison with the dimensions of the dispersion, and so the mean resultant force exerted on the particles inside τ, that is, by stress, -S say, which is a function of the local particle concentration.
>
> "Electostatic interparticle forces are conservative, and in that case one can interpret -S as the derivative of the mean potential energy per particle with respect to the volume of the mixture per particle. The contribution to the net force exerted on particles in our control volume by external particles is then
>
> $$-A \int_{x_1}^{x_2} \frac{\partial S}{\partial x} dx.$$
>
> A repulsive force between particles corresponds to a positive value of S (relative to zero when the particles are far apart), in which case S plays a dynamical role analogous to the pressure in a gas."

The equations of motion (23) and (24) are now reduced to

$$\frac{\partial \phi}{\partial t} + \frac{\partial}{\partial z} \phi u_p = 0, \tag{37}$$

$$\rho_p \phi \left[\frac{\partial u_p}{\partial t} + u_p \frac{\partial u_p}{\partial z} \right] = \mathcal{F} - \frac{2}{3} d \frac{\partial \mathcal{F}}{\partial z} \tag{38}$$

where $\mathcal{F}$ is given by (35) and

$$\frac{\partial \mathcal{F}}{\partial z} = \frac{\partial \mathcal{F}}{\partial \varepsilon} \frac{\partial \varepsilon}{\partial z} + \frac{\partial \mathcal{F}}{\partial u_p} \frac{\partial u_p}{\partial z} \tag{39}$$

and

$$\frac{\partial \mathfrak{F}}{\partial \varepsilon} = \frac{\partial N}{\partial \varepsilon}F + N\frac{\partial F}{\partial \varepsilon}\,. \tag{40}$$

Equations (37) and (38) are two nonlinear equations in two unknowns, ε and u_p. These equations differ from the ones derived by Foscolo and Gibilaro [1984] to which they reduce when the two additional terms

$$\frac{\partial \mathfrak{F}}{\partial u_p}\frac{\partial u_p}{\partial z} \tag{41}$$

and

$$\frac{\partial N}{\partial \varepsilon}\mathfrak{F} \tag{42}$$

are put to zero. The term (42) vanishes in the analysis of stability of uniform fluidization but (41) does not.

The term (41) also is absent from the list of forces which act in this problem developed by Batchelor. Hence, we are obliged to consider the physical origin of such term. We may regard the term (41) as arising from changes in the microstructure of the mixture. This has been well expressed in a recent paper by Ham and Homsy [1988].

> "Analysis of the mean settling speed leaves unresolved the problem of microstructural evolution in suspensions. Such changes in the relative positions of particles are likely because each particle in a random suspension sees a slightly different local environment and is therefore expected to have a velocity which is, in general different from that of any neighboring particle. The variations in particle velocities will lead to an adjustment of the particle distribution."
>
> They note further that
>
> "...the microstructural dependence arises from the fact that the time between the velocities of the faster-and slower-setting particles, and the difference will be influenced by the relative position of the particles. The influence of ϕ comes about from the change in interparticle spacing with concentration of particles."

8. Classification of Type and Hadamard Instability.

The theory of classification of type of a second order partial differential equation

$$\widehat{A}\frac{\partial^2 \phi}{\partial t^2} + \widehat{B}\frac{\partial^2 \phi}{\partial t \partial z} + \widehat{C}\frac{\partial^2 \phi}{\partial z^2} + \text{ lower order terms } = 0 \tag{43}$$

is well known. Everything depends on the discriminant

$$D = \widehat{B}^2 - 4\widehat{A}\widehat{C}.$$

Equation (43) is parabolic, elliptic or hyperbolic depending on whether the discriminant $D = 0$, $D < 0$ or $D > 0$, respectively. Its characteristics are given by

$$\frac{dz}{dt} = \frac{\hat{B} \pm \sqrt{\hat{B}^2 - 4\hat{A}\hat{C}}}{2\,\hat{A}} .$$

Hadamard instability is an explosive instability to short waves, the growth rates of unstable disturbances tend to infinity with α, that is the wave length $\frac{2\pi}{\alpha} -> 0$. Problems which are Hadamard unstable are ill-posed as initial value problems. In the analysis of short waves lower order terms are unimportant because the highest order derivatives dominate and the coefficients of these derivatives can not vary much in the length of short wave. This gives rise to the second order equation (43) with no lower order terms and constant coefficients. To show that the initial value problem is Hadamard unstable when it is elliptic we use normal modes

$$\phi(z,t) = P\, e^{i\alpha(z-\omega t)}$$

in equation (43) and obtain

$$\hat{A}(\alpha\omega)^2 - \hat{B}\alpha^2\omega + \hat{C}\alpha^2 = 0.$$

Hence

$$\alpha\omega = \alpha\left[\frac{\hat{B} \pm \sqrt{\hat{B}^2 - 4\hat{A}\hat{C}}}{2\,\hat{A}}\right].$$

Then the growth rate is

$$\sigma = Im[\alpha\omega] = \alpha Im\left[\frac{\hat{B} \pm \sqrt{\hat{B}^2 - 4\hat{A}\hat{C}}}{2\,\hat{A}}\right] = \pm\frac{\alpha}{2\,\hat{A}}\, Im[\sqrt{D}]$$

where $Im[.]$ stands for the imaginary part. Clearly, if $D > 0$ then $\sigma = 0$ so the problem is not Hadamard unstable. but if $D < 0$ then $\sigma -> \pm\infty$ as $\alpha -> \infty$, i.e. if a problem is elliptic it is also Hadamard unstable in the sense of an initial value problem.

9. Stability of Uniform Fluidization. In a state of uniform fluidization $\phi = \phi_0$ and $U = U(\phi_0) = U_0$ are constant and $u_p = 0$. Equations (2) and (22) are satisfied. Since $F = F_0 = 0$ for uniform fluidization and π_p is constant, (23) and (25) are satisfied. The mass balance equation (16) for balls of radius R is also satisfied. The state of uniform fluidization satisfies all the required equations.

Now we linearize the equations around the state of uniform fluidization. Let ϕ be the perturbation of ϕ_0, $N(z+x,t)$ be the perturbation of the uniform number density N_0 and u_p be the perturbation of u_p from zero. The mass balance equation (2) becomes

$$\frac{\partial\phi}{\partial t} + \phi_0\frac{\partial u_p}{\partial z} = 0 . \tag{44}$$

The mass balance equation (16) becomes

$$(45)\quad \frac{\partial}{\partial t}\int_{-R}^{R} N(z+x,t)\,(R^2-x^2)dx + N_0\,\frac{\partial}{\partial z}\int_{-R}^{R} u_p(z+x,t)(R^2-x^2)dx = 0\,.$$

The momentum equation (22) becomes

$$(46)\qquad \phi_0(1+\theta)\frac{\partial u_p}{\partial t} = -Q_0\frac{\partial \phi}{\partial z} + \frac{\gamma\tilde{g}}{U_0}\phi 0\left(\frac{\partial U}{\partial \phi_0}\,\phi + u_p\right) + \frac{\partial\left(\phi_0\eta_0\dfrac{\partial u_p}{\partial z}\right)}{\partial z}\,.$$

The linearization of (38) becomes

$$(47)\qquad \phi_0\frac{\partial u_p}{\partial t} = -B(\phi_0 u_p - C_1\phi) - C_2^2\frac{\partial \phi}{\partial z} + \frac{2}{3}\phi_0 d\,B\frac{\partial u_p}{\partial z}$$

where $B = \dfrac{4.8\tilde{g}\,(1-\phi_0)}{nu_c}$, $C_1 = \dfrac{n\,u_c\phi_0}{1-\phi_0}$ and $C_2^2 = 3.2\phi_0\tilde{g}d$.
After eliminating u_p between (44) and (47) we get the following second order equation

$$(48)\qquad \frac{\partial^2\phi}{\partial t^2} - \frac{2}{3}\,d\,B\frac{\partial^2\phi}{\partial t\partial z} - C_2^2\frac{\partial^2\phi}{\partial z^2} + B\left(\frac{\partial\phi}{\partial t} + C_1\frac{\partial\phi}{\partial z}\right) = 0\,.$$

Similarly, (45) and (47) give the following second order equation

$$(49)\qquad \begin{aligned}&\frac{\partial^2 u_p}{\partial t^2} - \frac{2}{3}\,d\,B\frac{\partial^2 u_p}{\partial t\partial z} - C_2^2\frac{\partial^2}{\partial z^2}\int_{-R}^{R} u_p(z+x,t)(R^2-x^2)dx\\ &\quad + B\left(\frac{\partial u_p}{\partial t} + C_1\frac{\partial}{\partial z}\int_{-R}^{R} u_p(z+x,t)\,(R^2-x^2)dx\right) = 0\,,\end{aligned}$$

and (44) and (46) give

$$(50)\qquad (1+\theta)\frac{\partial^2 u_p}{\partial t^2} = Q_0\frac{\partial^2 u_p}{\partial z^2} + \frac{\gamma\tilde{g}}{U_0}\left(-\phi_0\frac{\partial U}{\partial \phi_0}\frac{\partial u_p}{\partial z} + \frac{\partial u_p}{\partial t}\right) + \eta_0\frac{\partial^3 u_p}{\partial t\partial z^2}\,.$$

For (48) D (see section 8) is

$$D = \left(\frac{2}{3}\,d\,B\right)^2 + 4C_2^2\,,$$

and the characteristics are given by

$$\frac{dz}{dt} = -\frac{1}{3}\,d\,B \pm \sqrt{\left(\frac{1}{3}\,d\,B\right)^2 + C_2^2}\,.$$

This shows that if B and C_2 are not both zero (there is a particle phase pressure) then the governing second order equation is hyperbolic and has two real characteristics. This means the initial value problem for this equation is well posed and will not give rise to Hadamard instability. In equation (49) the second order term with an integral can be considered to be of a lower order and hence the expression for D becomes

$$D = \left(\frac{2}{3}d\ B\right)^2 ,$$

and the characteristics are given by

$$\frac{dz}{dt} = 0,\ -\frac{2}{3}\ d\ B\ .$$

If the third order term in (50) is dropped then D and the characteristics are given by

$$D = \sqrt{4\ Q_0\ (1+\theta)}\ ,$$

$$\frac{dz}{dt} = \pm\sqrt{\frac{Q_0}{1+\theta}}\ .$$

We next look for normal mode solutions of (48), (49) and (50)

$$\phi(z,t) = P\ e^{i\alpha(z-\omega t)},\quad u_p(z,t) = Q\ e^{i\alpha(z-\omega t)}$$

where $i = \sqrt{-1}$, P and Q are constants, α is the wave number and $\alpha\omega$ is the angular frequency. By putting the above solutions in (48), (49) and (50) we get following dispersion relations

(51) $$(\alpha\omega)^2 + \frac{2}{3}d\ B\alpha^2\omega - C_2^2\alpha^2 + B(i\alpha\omega - C_1\ i\alpha) = 0\ ,$$

(52) $$(\alpha\omega)^2 + \frac{2}{3}d\ B\alpha^2\omega - C_2^2\alpha^2\Theta(\alpha) + B(i\alpha\omega - C_1\ i\alpha\Theta(\alpha)) = 0\ ,$$

(53) $$(1+\theta)(\alpha\omega)^2 = Q_0\alpha^2 + \frac{\gamma\tilde{g}}{U_0}\left(\phi_0\frac{\partial U}{\partial\phi_0}\quad i\alpha + i\alpha\omega\right) - \eta_0\ \alpha^3\omega i,$$

where $\Theta(\alpha) = 3\left[\frac{\sin\ \alpha R}{(\alpha R)^3} - \frac{\cos\ \alpha R}{\alpha R)^2}\right]$ (see figure 2).

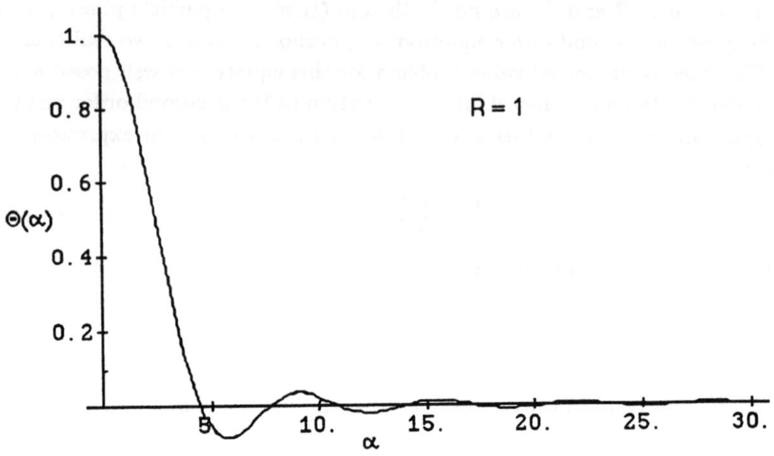

FIGURE 2. For $R = 1$, $\Theta(\alpha)$ is plotted as a function of α.

From (51) we find that the system is unstable if

$$C_1 + \frac{1}{3} d\, B > \sqrt{C_2^2 + \left(\frac{1}{3}\, d\, B\right)^2}. \tag{54}$$

This reduces to the well known criterion of Wallis when $\dfrac{\partial^2 \phi}{\partial t \partial z}$ term in (48) is dropped. From (52) the bed is unstable if

$$\Theta(\alpha) C_1 + \frac{1}{3}\, d\, B > \sqrt{\Theta(\alpha) C_2^2 + \left(\frac{1}{3}\, d\, B\right)^2}. \tag{55}$$

This also reduces to a form very similar to the well known criterion of Wallis when $\dfrac{\partial^2 \phi}{\partial t \partial z}$ term in (49) is dropped. In the present case, the inequalities (54) and (55) are always satisfied and so the uniform state is always unstable. From equation (53) the bed is unstable if

$$\frac{\dfrac{\gamma \tilde{g}}{U_0} \phi_0 \dfrac{\partial U}{\partial \phi_0}}{\dfrac{\gamma \tilde{g}}{U_0} + \eta_0\, \alpha^2} > \sqrt{\frac{Q_0}{1+\theta}}. \tag{56}$$

Conditions (55) and (56) are different because of their dependence on the wave number. The wave number dependence of (55) makes the modes with $\Theta(\alpha) = 0$ neutrally stable. Relation (56) shows that Batchelor's equations are stable to short waves and there exists a lower bound on α above which all modes are stable.

Viscosity stabilizes short waves. We want readers to notice that the criterion for stability in the absence of finite size of particles or viscosity effects is independent of the wave number (cf. Jones and Prosperetti [1985], Prosperetti and Jones [1987] and Prosperetti and Satrape [1989]).

Now consider the case when in (52) terms coming from the particle phase pressure are dropped. We solve for $\alpha\omega$

$$\alpha\omega = \frac{1}{2}\left(-Bi \pm \sqrt{-B^2 + 4\ B\ C_1\ i\alpha\ \theta}\right)$$

$$= \frac{B\ i}{2}\left(-1 \pm \sqrt{1 - \frac{4\ C_1\ i\alpha\ \Theta}{B}}\right)$$

(57) $$= \frac{B\ i}{2}\left(-1 \pm \sqrt{1 - \Sigma\ i}\right)$$

where $\Sigma = \dfrac{4\ C_1\ \alpha\Theta}{B}$.

For $|\Sigma| < 1$ we can take Taylor series expansion

$$\alpha\omega = \frac{B\ i}{2}\left(-1 \pm \left(1 - \frac{\Sigma i}{2} + \frac{\Sigma^2}{8} + \ldots\right)\right) .$$

The growth rate

$$\sigma = Im[\alpha\omega] = \frac{B}{2}\left(-1 \pm \left(1 + \frac{\Sigma^2}{8} + \ldots\right)\right)$$

where $Im[\alpha\omega]$ is the imaginary part of $\alpha\omega$. The above expression for the growth rate and (57) implies

(1) Instability is weak when Σ is small.

(2) Σ is a bounded function of α with $\Sigma - > 0$ as $\alpha - > 0$ or $\alpha - > \infty$ (see figure 3). Thus both short and long waves are neutrally stable.

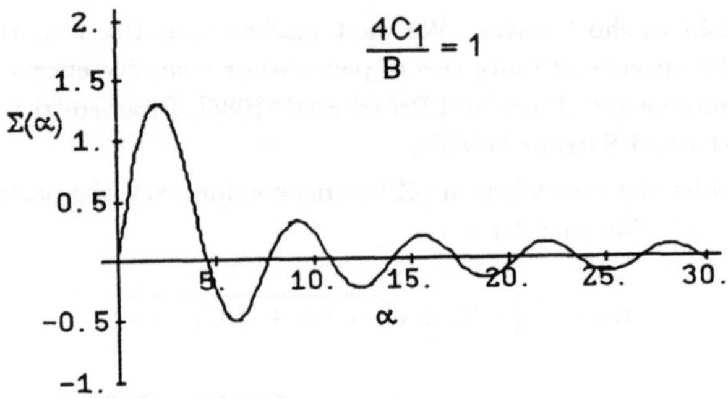

FIGURE 3. For $\dfrac{4\ C_1}{B} = 1$ and $R = 1$, $\Sigma(\alpha)$ is plotted as a function of α.

(3) Σ is maximum for $\alpha_{\max} \approx \dfrac{2.1}{R}$ and $\Sigma_{\max} \approx \dfrac{4\ C_1(1.3)}{B\ R}$ (see figure 3). Note that both $\alpha_{\max}$ and $\Sigma_{\max}$ increase as R is decreased. Hence the problem becomes ill-posed as $\alpha -> \infty$ for $R = 0$ (because $\Sigma_{\max} -> \infty$). To see that the problem is Hadamard unstable we put $R = 0$ in equation (57) and find that

(58) $$\alpha\omega = \frac{B\ i}{2}\left(-1 \pm \sqrt{1 - \frac{4\ C_1\ \alpha}{B}i}\right)$$

where we have used $\lim\limits_{R->0} \Theta = 1$. Therefore $\sigma = Im[\alpha\omega] = c\sqrt{\alpha}$ for some constant c, independent of α, for large α. The growth rate is unbounded for large α. Hence, the uniform fluidization is Hadamard unstable. Consider again equation (49) which reduces to the following form in the present case

(59) $$\frac{\partial^2\phi}{\partial t^2} + B\left(\frac{\partial\phi}{\partial t} + C_1\frac{\partial\phi}{\partial z}\right) = 0.$$

This equation is parabolic because $D = 0$. So Hadamard instability here does not arise because the characteristics are imaginary. This kind of Hadamard instability is similar to that of the backward heat equation, here with the roles of time and space interchanged. One can see this by looking for spatially growing modes

(60) $$\phi(z,t) = P\ e^{i\alpha t}e^{\sigma z}$$

where σ is complex and α is real. After combining (59) and (60), we get

$$\sigma = \frac{\alpha^2}{C_1} - \frac{Bi\alpha}{C_1}\ .$$

This is the kind of dispersion one get for the Hadamard instability of the backward heat equation.

(4) If we put the values of C_1 and B in the expression for Σ we get

$$\Sigma_{\max} \approx 6.9\frac{C_1^2}{C_2^2} .$$

So if $C_2 > \sqrt{6.9}\, C_1$ then Σ is smaller than one. In this case the instability is expected to be weak. This may be compared with the criterion $C_2 > C_1$ for stability derived in the (1984) paper of Foscolo and Gibilaro.

REFERENCES

BATCHELOR, G.K., *A new theory of the instability of a uniform fluidized bed*, J. Fluid Mech. 193 (1988), 75–110.

DREW, D., *Mathematical modeling of two-phase flow*, A. Rev. Fluid Mech. 15 (1983), 261–291.

FOSCOLO, P.V. AND GIBILARO, L.G., *A fully predictive criterion for transition between particulate and aggregate fluidization*, Chem. Eng. Sci. 39 (1984), 1667–1674.

FOSCOLO, P.V. AND GIBILARO, L.G., *Fluid dynamic stability of fluidized suspensions. The particle bed model*, Chem Eng Sci. 42 (1987), 1489–1500.

HAM, J.M. AND HOMSY, G.M., *Hindered settling and hydrodynamic dispersion in quiescent sedimenting suspensions*, Int. J. Multiphase Flow 14 (1988), 533–546.

JONES, A.V. AND PROSPERETTI, A., *On the suitability of first-order differential models for two-phase flow prediction*, Int. J. Multiphase Flow 11 (1985), 133–148.

JOSEPH, D.D., *Generalization of Foscolo-Gibilaro analysis of dynamic waves*, Chem Eng Sci., 45 (1990), 411–414.

JOSEPH, D.D. AND LUNDGREN, T.S., *Ensemble averaged and mixture theory equations for incompressible fluid-particle suspensions*, Int. J. Multiphase Flow, 16 (1990), 35–42.

PROSPERETTI, A. AND JONES, A.V., *The linear stability of general two-phase flows models- II*, Int. J. Multiphase Flow 13 (1987), 161–171.

PROSPERETTI, A. AND SATRAPE, J., *Stability of two-phase flows models*, Proceedings of the January, 1989 workshop of the IMA on "Two phase flows in fluidized beds, sedimentation and granular flow". To be published by Springer-Verlag (1989).

SINGH, P., CAUSSIGNAC, PH., FORTES, A., JOSEPH D.D. AND LUNDGREN, T., *Stability of periodic arrays of cylinders across the stream by direct simulation*, J. Fluid Mech. 205 (1989), 553–571.

ON GEURST'S EQUATIONS FOR INERTIAL COUPLING IN TWO-PHASE FLOW*

GRAHAM B. WALLIS†

Abstract. Geurst used variational techniques to derive the inertia terms in the momentum equations of the two fluid model for two-phase flow.

This paper compares Geurst's results with other formulations, examines the significance of various terms, and explores resulting predictions in one-dimensional steady and unsteady flow situations.

Introduction. In a series of papers [1,2,3], Geurst has used variational techniques to derive various forms of momentum conservation equations for two-phase flow. His basic assumption is that the kinetic energy density is

$$k = \frac{1}{2}\rho_1\alpha_1 v_1^2 + \frac{1}{2}\rho_2\alpha_2 v_2^2 + \frac{1}{2}\rho_1 m(v_1 - v_2)^2 \tag{1}$$

where ρ_1 and ρ_2 are the densities of the individual phases, α_1 and α_2 their volumetric concentration, v_1 and v_2 their average velocities and "m" is a coefficient dependent on α_1 (or $\alpha_2 = 1 - \alpha_1$). In essence, (1) postulates additional kinetic energy due to the flow field set up by the relative motion.

Using an alternative approach, Wallis [4] derived an expression similar to (1) with m expressed in a different nomenclature

$$m = \alpha_1(\alpha_1\beta - 1) \tag{2}$$

The above parameters are related [4] to the "virtual mass coefficient" used by Drew and Lahey [6]:

$$c_{vm} = (\alpha_1\beta - 1)\frac{\alpha_1}{\alpha_2} = \frac{m}{\alpha_2} \tag{3}$$

Several comparisons between Geurst's theory and other formulations have already been made [4,7], with a few simple tests apparently favoring his predictions. The purpose of the present note is both to extend these comparisons in more general terms and to derive specific results for cases of interest. The nomenclature chosen

*Prepared for the Institute for Mathematics and its Applications workshop "Two Phase Waves in Fluidized Beds, Flowing Composites and Granular Media," University of Minnesota, January, 1989

†Professor, Thayer School of Engineering, Dartmouth College, Hanover, New Hampshire 03755, USA

is that defined in [4] and differs from Geurst's, the latter defining "density" of each phase as mass per unit total volume rather than per unit volume of that phase.

Forms of the Equations of Motion. A useful starting point is the equations of momentum conservation of each phase presented as equations (5.7) and (5.8) of [3]. In the present notation these are

$$\frac{\partial}{\partial t}(\rho_1\alpha_1\mathbf{v}_1) + \nabla\cdot\{\rho_1\alpha_1\mathbf{v}_1\mathbf{v}_1 + \rho_1 m(\mathbf{v}_2-\mathbf{v}_1)(\mathbf{v}_2-\mathbf{v}_1)\}$$

$$+\alpha_1\nabla p_2 + \nabla\cdot\left\{\frac{1}{2}\rho_1(m+\alpha_1 m')(\mathbf{v}_2-\mathbf{v}_1)^2\right\} = \mathbf{M}_1^d + \alpha_1\mathbf{f}_1 \tag{4}$$

$$\frac{\partial}{\partial t}(\rho_2\alpha_2\mathbf{v}_2) + \nabla\cdot(\rho_2\alpha_2\mathbf{v}_2\mathbf{v}_2) + \alpha_2\nabla p_2 = -\mathbf{M}_1^d + \alpha_2\mathbf{f}_2 \tag{5}$$

Here m' denotes $dm/d\alpha_2$. $\mathbf{f}_1$ and $\mathbf{f}_2$ have been added to account for any additional forces on the phases, per unit volume of each phase. $\mathbf{M}_1^d$ is a mutual force that was derived by Geurst [3] in his (5.9) as

$$\mathbf{M}_1^d = \frac{\partial}{\partial t}\{\rho_1 m(\mathbf{v}_2-\mathbf{v}_1)\} + \nabla\cdot\{\rho_1 m\mathbf{v}_2(\mathbf{v}_2-\mathbf{v}_1)\} + \rho_1 m(\mathbf{v}_2-\mathbf{v}_1)\cdot(\nabla\mathbf{v}_2)^T \tag{6}$$

As a first simple conclusion, we note that when the dispersed phase is stationary ($\mathbf{v}_2 = 0$) and the flow is steady, $\mathbf{M}_1^d$ disappears and (5) reduces to

$$\nabla p_2 = \mathbf{f}_2 \tag{7}$$

which is the same result derived for these conditions in (3.214) of [4]. If the flow pattern is known, ∇p_2 can be computed from the velocity field using (4), allowing $\mathbf{f}_2$ to be determined.

We now turn to the terms on the left-hand side of (4) and (5). From (2) it follows that

$$m + \alpha_1 m' = -\alpha_1^2\frac{d}{d\alpha_1}(\alpha_1\beta - 1) \tag{8}$$

Now, in [4] the mean pressure in the continuous phase 1 was shown to be related to the mean pressure in the discontinuous phase by

$$p_1 = p_2 + \xi\rho_1 w^2 \tag{9}$$

where w^2 is the square of the magnitude of the relative velocity and

$$\xi = -\frac{\alpha_1}{2}\frac{d(\alpha_1\beta - 1)}{d\alpha_1} \tag{10}$$

Combining (8), (9) and (10) we find

$$p_1 = p_2 + \frac{m + \alpha_1 m'}{2\alpha_1}\rho_1 w^2 \tag{11}$$

In all of this, w^2 is to be interpreted as the scalar part of the dyadic $(\mathbf{v}_2 - \mathbf{v}_1)^2$, i.e.

$$(\mathbf{v}_2 - \mathbf{v}_1)^2 = \mathbf{I}w^2 \tag{12}$$

Using (11) and (12) we have

$$\nabla \cdot \frac{1}{2}\rho_1(m + \alpha_1 m')(\mathbf{v}_2 - \mathbf{v}_1)^2 = \nabla\left[\alpha_1(p_1 - p_2)\right] \tag{13}$$

Making this substitution in (4) the pressure terms combine to two terms, $\alpha_1\nabla p_1 + (p_1 - p_2)\nabla\alpha_1$, that have previously appeared in some versions of the "two-fluid model," e.g. Pauchon-Banerjee [8].

If we now add together (4) and (5), with the above conclusion incorporated, the result is

$$\frac{\partial}{\partial t}(\rho_1\alpha_1\mathbf{v}_1 + \rho_2\alpha_2\mathbf{v}_2) + \nabla\cdot\left[\rho_1\alpha_1\mathbf{v}_1\mathbf{v}_1 + \rho_2\alpha_2\mathbf{v}_2\mathbf{v}_2 + \rho_1 m(\mathbf{v}_2 - \mathbf{v}_1)(\mathbf{v}_2 - \mathbf{v}_1)\right.$$

$$\left. + (\alpha_1 p_1 + \alpha_2 p_2)\mathbf{I}\right] = \alpha_1\mathbf{f}_1 + \alpha_2\mathbf{f}_2 \tag{14}$$

where use was made of the constraint

$$\alpha_1 + \alpha_2 = 1 \tag{15}$$

Eq. (14) is exactly what would be expected as the overall momentum conservation equation for the mixture, the term in square brackets being the combined pressure and momentum flux tensor given as (3.152) in [4].

The most interesting term is perhaps the "phase interaction" defined in (6). As written, the expression does not obviously relate to the more familiar of the inertial coupling formulations in the literature. However, this may be corrected with a little analysis.

For a start, the last term in (6) may be expressed as

$$\rho_1 m(\mathbf{v}_2 - \mathbf{v}_1)\cdot(\nabla \mathbf{v}_2)^T = \rho_1 m(\mathbf{v}_2 - \mathbf{v}_1)\cdot \nabla \mathbf{v}_2 + \rho_1 m(\mathbf{v}_2 - \mathbf{v}_1)\times\nabla\times\mathbf{v}_2 \tag{16}$$

The last term in (16) differs from the "lift force" derived by Drew and Lahey [6] in that it contains the curl of the average velocity of the discontinuous phase rather than the continuous phase. To change to the latter we use the identity

$$(\mathbf{v}_2 - \mathbf{v}_1)\times\nabla\times(\mathbf{v}_2 - \mathbf{v}_1) = \nabla\frac{(\mathbf{v}_2 - \mathbf{v}_1)^2}{2} - (\mathbf{v}_2 - \mathbf{v}_1)\cdot\nabla(\mathbf{v}_2 - \mathbf{v}_1) \tag{17}$$

Expanding the other terms in (6) and using (16) and (17) we obtain

$$\mathbf{M}_1^d = \rho_1 m\left[\left(\frac{\partial}{\partial t} + \mathbf{v}_2\cdot\nabla\right)\mathbf{v}_2 - \left(\frac{\partial}{\partial t} + \mathbf{v}_1\cdot\nabla\right)\mathbf{v}_1 + (\mathbf{v}_2 - \mathbf{v}_1)\times\nabla\times\mathbf{v}_1\right]$$

$$+\rho_1 m\nabla\frac{(\mathbf{v}_2 - \mathbf{v}_1)^2}{2} + (\mathbf{v}_2 - \mathbf{v}_1)\left[\frac{\partial}{\partial t}(\rho_1 m) + \nabla\cdot(\rho_1 m\mathbf{v}_2)\right] \tag{18}$$

The first term in square brackets is the same as the combination of "virtual mass" and "lift" forces deduced by Drew and Lahey [6]. The second term is of the "Bernoulli" type [4,5]. The final term involves a total rate of change of the inertial coupling coefficient itself, which is not an unreasonable inclusion since this is one way in which the properties of the assembly can be changed.

Indeed, in an accelerating suspension in which all of the properties are uniform, $\mathbf{M}_1^d$ reduces to $d/dt\,\{\rho_1 m(\mathbf{v}_2 - \mathbf{v}_1\}$ and the equations of motion become, in the case where external forces act only on the particles,

$$\frac{d}{dt}(\rho_1\alpha_1\mathbf{v}_1) + \alpha_1\nabla p_2 = \frac{d}{dt}\{\rho_1 m(\mathbf{v}_2 - \mathbf{v}_1)\} \tag{19}$$

$$\frac{d}{dt}(\rho_2\alpha_2\mathbf{v}_2) + \alpha_2\nabla p_2 = -\frac{d}{dt}\{\rho_1 m(\mathbf{v}_2 - \mathbf{v}_1)\} + \alpha_2\mathbf{f}_2 \tag{20}$$

We now consider a very dilute uniform suspension of essentially independent particles of volume V_2 each of which has identical motion. If the fluid is at rest and not accelerating, there is negligible overall pressure gradient. The equation of motion of each particle under an applied force $\mathbf{F}_2$ is

$$\frac{d}{dt}[(\rho_2 V_2 \mathbf{v}_2) + c_{vm}\rho_1 V_2 \mathbf{v}_2] = \mathbf{F}_2 \tag{21}$$

which agrees with (20) if we divide by the total volume V (per particle) and interpret the added mass coefficient, as in (3), as

$$c_{vm} = \frac{m}{\alpha_2} \tag{22}$$

If the fluid is in addition given an acceleration, there will be set up a mean pressure gradient such that

$$\rho_1 \frac{d\mathbf{v}_1}{dt} + \nabla p = 0 \tag{23}$$

Eq. (23) is compatible with (19) because, by continuity

$$\frac{d}{dt}(\rho_1 \alpha_1) = 0 \tag{24}$$

and the apparent mass term is vanishingly small. It does not matter which pressure gradient is used since all pressures at corresponding points in each "unit cell," differ from each other by a constant throughout the suspension. When uniform fluid motion is superimposed on the motion represented by (21), it does not change ***additional*** fluid momentum associated with each particle, which depends only on the relative velocity. The uniform acceleration of phase 1 merely adds an effective body force field experienced by both phases. Then (21) becomes

$$\frac{d}{dt}[(\rho_2 V_2 \mathbf{v}_2) + c_{vm}\rho_1 V_2(\mathbf{v}_2 - \mathbf{v}_1)] - V_2 \rho_1 \frac{d\mathbf{v}_1}{dt} = \mathbf{F}_2 \tag{25}$$

If we divide (25) by the total volume V (per particle), the result, in view of (23), is (20). Therefore, there appear to be valid physical arguments for including a term such as the final one in (18).

The question now arises whether any insights are to be gained by combining terms from the left- and right-hand sides of (4) in any particular fashion. Clearly the continuity equations:

$$\frac{\partial}{\partial t}(\alpha_1 \rho_1) + \nabla \cdot (\rho_1 \alpha_1 \mathbf{v}_1) = 0 \tag{26}$$

$$\frac{\partial}{\partial t}(\alpha_2\rho_2) + \nabla\cdot(\rho_2\alpha_2\mathbf{v}_2) = 0 \tag{27}$$

can be used to express the two lead terms of (4) and (5) as density times acceleration, in the usual way. Geurst, in (6.2) and (6.3) of [3], makes several combinations (there seem to be some errors in his (6.2)) without interpreting them physically.

Rather than tackling the whole equation we first consider a simple case of *steady flow*, with no additional force $\mathbf{f}_1$, through a stationary particle lattice ($\mathbf{v}_2 = 0$). $\mathbf{M}_1^d$ is thus zero, from (6), and we may use (2), (8), (13), and (26) in (4) to get

$$\alpha_1\rho_1\mathbf{v}_1\cdot\nabla\mathbf{v}_1 + \alpha_1\nabla p_1 - \frac{\alpha_1}{2}\rho_1 v_1^2\nabla(\alpha_1\beta) + \nabla\cdot[\rho_1\alpha_1(\alpha_1\beta-1)\mathbf{v}_1\mathbf{v}_1] = 0 \tag{28}$$

The final term in (28) may be expanded:

$$\nabla\cdot[(\rho_1\alpha_1\mathbf{v}_1)(\alpha_1\beta-1)\mathbf{v}_1] = (\alpha_1\beta-1)\mathbf{v}_1\nabla\cdot(\rho_1\alpha_1\mathbf{v}_1) + \rho_1\alpha_1\mathbf{v}_1\cdot\nabla[(\alpha_1\beta-1)\mathbf{v}_1] \tag{29}$$

Using (29) and (26) in (28) we obtain

$$\frac{1}{\rho_1}\nabla p_1 - \frac{v_1^2}{2}\nabla(\alpha_1\beta) + \mathbf{v}_1\cdot\nabla(\alpha_1\beta\mathbf{v}_1) = 0 \tag{30}$$

Now, the final term in (30) may be expanded as

$$\mathbf{v}_1\cdot\nabla(\alpha_1\beta\mathbf{v}_1) = \nabla(\alpha_1\beta v_1^2) - \alpha_1\beta\mathbf{v}_1\cdot\nabla\mathbf{v}_1.$$

$$-\mathbf{v}_1\times\nabla\times(\alpha_1\beta\mathbf{v}_1) - \alpha_1\beta\mathbf{v}_1\times\nabla\times\mathbf{v}_1 \tag{31}$$

As discussed in [4], the fluid flow has a scalar "potential," ϕ, such that

$$-\nabla\phi = \alpha_1\beta\mathbf{v}_1 \tag{32}$$

Therefore

$$\nabla\times(\alpha_1\beta\mathbf{v}_1) = 0 \tag{33}$$

We have, in addition, the identity

$$\nabla(\alpha_1\beta v_1^2) = \alpha_1\beta\nabla v_1^2 + v_1^2\nabla(\alpha_1\beta) \tag{34}$$

Using (31), (33) and (34) in (30) we obtain

$$\frac{1}{\rho_1}\nabla p_1 + \nabla\left(\frac{1}{2}\alpha_1\beta v_1^2\right) = 0 \tag{35}$$

which is the ***averaged Bernoulli Equation*** for the continuous phase in a more general form than derived previously [4], being now valid for compressible flow.

To obtain a more general result, that is valid when the dispersed phase is moving, we return to (4), set $\mathbf{f}_1 = 0$, use (8) and (26), (6), (16) and (17), expand the time derivative on the right-hand side and combine the divergence terms to get

$$\rho_1\alpha_1\left(\frac{\partial \mathbf{v}_1}{\partial t} + \mathbf{v}_1\cdot\nabla\mathbf{v}_1\right) - \nabla\cdot[\rho_1\alpha_1(\alpha_1\beta-1)\mathbf{v}_1(\mathbf{v}_2-\mathbf{v}_1)] - \frac{\alpha_1\rho_1}{2}(v_2-v_1)^2\nabla(\alpha_1\beta-1)$$

$$+\alpha_1\nabla p_1 = \rho_1\alpha_1\frac{\partial}{\partial t}[(\alpha_1\beta-1)(\mathbf{v}_2-\mathbf{v}_1)] + (\alpha_1\beta-1)(\mathbf{v}_2-\mathbf{v}_1)\frac{\partial}{\partial t}(\rho_1\alpha_1)$$

$$+\rho_1\alpha_1(\alpha_1\beta-1)\left[(\mathbf{v}_2-\mathbf{v}_1)\cdot\nabla\mathbf{v}_1 + \nabla\frac{(v_2-v_1)^2}{2} + (\mathbf{v}_2-\mathbf{v}_1)\times\nabla\times\mathbf{v}_1\right] \tag{36}$$

We now make use of the velocity potential developed by Geurst [1,2,3] and Wallis [4] in the form

$$-\nabla\phi = \mathbf{v}_1 - (\alpha_1\beta-1)(\mathbf{v}_2-\mathbf{v}_1) \tag{37}$$

whence

$$\nabla\times\mathbf{v}_1 - \nabla\times[(\alpha_1\beta-1)(\mathbf{v}_2-\mathbf{v}_1)] = 0 \tag{38}$$

moreover,

$$\mathbf{v}_1\cdot\nabla\mathbf{v}_1 = \nabla\frac{v_1^2}{2} - \mathbf{v}_1\times\nabla\times\mathbf{v}_1 \tag{39}$$

and

$$\nabla \cdot [\rho_1\alpha_1(\alpha_1\beta - 1)\mathbf{v}_1(\mathbf{v}_2 - \mathbf{v}_1)] = (\alpha_1\beta - 1)(\mathbf{v}_2 - \mathbf{v}_1)\nabla \cdot (\rho_1\alpha_1\mathbf{v}_1)$$

$$+\rho_1\alpha_1\mathbf{v}_1 \cdot \nabla [(\alpha_1\beta - 1)(\mathbf{v}_2 - \mathbf{v}_1)] \tag{40}$$

Using all of (37) through (40) in (36) and again invoking (26) we find that the factor $\rho_1\alpha_1$ can be removed and we obtain

$$\frac{\partial}{\partial t}(-\nabla\phi) + \nabla\frac{v_1^2}{2} - \nabla\left[(\alpha_1\beta - 1)\frac{(v_2 - v_1)^2}{2}\right] + \frac{\nabla p_1}{\rho_1} = \mathbf{v}_1 \cdot \nabla [(\alpha_1\beta - 1)(\mathbf{v}_2 - \mathbf{v}_1)]$$

$$+(\alpha_1\beta - 1)[(\mathbf{v}_2 - \mathbf{v}_1) \cdot \nabla\mathbf{v}_1 + (\mathbf{v}_2 - \mathbf{v}_1)\times\nabla\times\mathbf{v}_1]$$

$$+\mathbf{v}_1\times\nabla\times [(\alpha_1\beta - 1)(\mathbf{v}_2 - \mathbf{v}_1)] \tag{41}$$

The terms on the right of (41) are the four components of $\nabla [(\alpha_1\beta - 1)(\mathbf{v}_2 - \mathbf{v}_1) \cdot \mathbf{v}_1)]$, and therefore (41) reduces to

$$\nabla\left[-\frac{\partial\phi}{\partial t} + \frac{v_1^2}{2} + \frac{(\alpha_1\beta - 1)}{2}(v_1^2 - v_2^2)\right] + \frac{\nabla p_1}{\rho_1} = 0 \tag{42}$$

This very simple result has the form of *Bernoulli's Equation for phase 1*, reducing to (35) when $\mathbf{v}_2 = 0$ in steady flow and to the corresponding single-phase expression when $\mathbf{v}_1 = \mathbf{v}_2$.

A similar development follows from (5) with $\mathbf{f}_2 = 0$. First we use the continuity equation and introduce factors $(\rho_2\alpha_2/\rho_2\alpha_2)$ to get

$$\rho_2\alpha_2\left(\frac{\partial\mathbf{v}_2}{\partial t} + \mathbf{v}_2 \cdot \nabla\mathbf{v}_2\right) + \alpha_2\nabla p_2 = -\frac{\partial}{\partial t}\left\{\rho_2\alpha_2\frac{\rho_1\alpha_1}{\rho_2\alpha_2}(\alpha_1\beta - 1)(\mathbf{v}_2 - \mathbf{v}_1)\right\}$$

$$-\nabla \cdot \left\{\rho_2\alpha_2\mathbf{v}_2\frac{\rho_1\alpha_1}{\rho_2\alpha_2}(\alpha_1\beta - 1)(\mathbf{v}_2 - \mathbf{v}_1)\right\}$$

$$-\rho_1\alpha_1(\alpha_1\beta - 1)[(\mathbf{v}_2 - \mathbf{v}_1) \cdot \nabla\mathbf{v}_2 + (\mathbf{v}_2 - \mathbf{v}_1)\times\nabla\times\mathbf{v}_2] \tag{43}$$

We now introduce the potential η introduced by Geurst [1,2,3] and Wallis [4],

$$-\nabla\eta = \mathbf{v}_2 + \frac{\alpha_1\rho_1}{\alpha_2\rho_2}(\alpha_1\beta - 1)(\mathbf{v}_2 - \mathbf{v}_1) \tag{44}$$

whence

$$\nabla\times\mathbf{v}_2 + \nabla\times\left[\frac{\alpha_1\rho_1}{\alpha_2\rho_2}(\alpha_1\beta - 1)(\mathbf{v}_2 - \mathbf{v}_1)\right] = 0 \tag{45}$$

Moreover, we have the identity

$$\mathbf{v}_2\cdot\nabla\mathbf{v}_2 = \nabla\frac{v_2^2}{2} - \mathbf{v}_2\times\nabla\times\mathbf{v}_2 \tag{46}$$

Expanding the various terms in (43) and using (44) to (46) as well as (27) yields, eventually

$$\nabla\left[-\frac{\partial\eta}{\partial t} + \frac{v_2^2}{2} + \frac{\alpha_1\rho_1}{\alpha_2\rho_2}(\alpha_1\beta - 1)(\mathbf{v}_2 - \mathbf{v}_1)\cdot\mathbf{v}_2\right] + \frac{\nabla p_2}{\rho_2} = 0 \tag{47}$$

which is *Bernoulli's Equation for phase 2.*

Eq. (42) and (47) are the three-dimensional equivalents of (3.14) and (3.15) in Geurst [1] where they are regarded as primitive, the equations of motion being derived from them, rather than the other way around, as we have done here. A point worth further attention is that (37) and (44) should only be valid for certain classes of *additional force fields*, $\mathbf{f}_1$ and $\mathbf{f}_2$, therefore, there are some restrictions on these developments.

For incompressible phases with a common stagnation pressure, p_0, we may integrate (42) and (47) and add $\alpha_1\rho_1$ times the former to $\alpha_2\rho_2$ times the latter to obtain

$$\alpha_1\rho_1\frac{\partial\phi_1}{\partial t} + \alpha_2\rho_2\frac{\partial\eta}{\partial t} + \frac{1}{2}\alpha_1\rho_1 v_1^2 + \frac{1}{2}\alpha_2\rho_2 v_2^2$$

$$+\frac{1}{2}\alpha_1\rho_1(\alpha_1\beta - 1)w^2 + \alpha_1 p_1 + \alpha_2 p_2 = p_0 \tag{48}$$

which is the combined Bernoulli Equation for the mixture and is related to (2.23) and (2.28) in Geurst [3] who points out that the terms involving velocity squared in (48) are exactly the same as the kinetic energy density in (1).

One-Dimensional Steady Flow. A simple application of (42) and (47) is the steady one-dimensional flow of incompressible components, as in a nozzle. Using (9), (10), and Maxwell's expression for the *exertia* [4],

$$\alpha_1\beta - 1 = \frac{\alpha_2}{2} \tag{49}$$

the pressures may be eliminated, in the case where the phases both come from a region at a common stagnation pressure, to obtain

$$\frac{1}{2}\rho_2 v_2^2 - \frac{1}{2}\rho_1 v_1^2 + \frac{1}{4}\rho_1(v_2^2 - v_1^2) = 0 \tag{50}$$

Therefore, the *slip ratio* is

$$\frac{v_2}{v_1} = \left(\frac{3\rho_1}{\rho_1 + 2\rho_2}\right)^{1/2} \tag{51}$$

which is the "minimum kinetic energy" solution given in [4] and is independent of void fraction.

More generally, if we do not assume (49) to be true, the simultaneous solution of (42) and (47) leads, with the use of (9) and (10), to

$$\rho_2\frac{v_2^2}{2} - \rho_1\frac{v_1^2}{2} + \frac{\alpha_1}{\alpha_2}\rho_1(\alpha_1\beta - 1)(v_2^2 - v_1 v_2) - \frac{\alpha_1\beta - 1}{2}\rho_1(v_1^2 - v_2^2)$$

$$+\frac{\alpha_1\rho_1}{2}(v_2 - v_1)^2\frac{d(\alpha_1\beta - 1)}{d\alpha_1} = 0 \tag{52}$$

In [4] it was found useful to define the parameter

$$B = \frac{\alpha_1\beta - 1}{\alpha_2} \tag{53}$$

which allows (52) to be transformed to

$$\rho_2\frac{v_2^2}{2} - \rho_1\frac{v_1^2}{2} + \alpha_1\rho_1 B(v_2^2 - v_1 v_2) - \frac{B}{2}\rho_1(1 - \alpha_1)(v_1^2 - v_2^2)$$

$$+\frac{\alpha_1\rho_1}{2}(v_2 - v_1)^2\left(-B + \alpha_2\frac{dB}{d\alpha_1}\right) = 0 \tag{54}$$

which may be rearranged to

$$-\rho_1 v_1^2(1+B) + \rho_2 v_2^2\left(1+\frac{\rho_1 B}{\rho_2}\right) + \rho_1\alpha_1\alpha_2(v_2-v_1)^2\frac{dB}{d\alpha_1} = 0 \tag{55}$$

Eq. (55) is exactly the condition given as (6.6) in [4] for the volumetric integral of the kinetic energy density to be minimized throughout the nozzle.

Since B is a function of α_1 and constant α_1 implies constant v_2/v_1, (55) may be solved for (v_2/v_1) treating $dB/d\alpha_1$ as constant. The condition for real roots turns out to be

$$\frac{dB}{d\alpha_1}\alpha_1\alpha_2(\rho_2-\rho_1) < (1+B)(\rho_2+\rho_1 B) \tag{56}$$

which is always satisfied by a Maxwellian suspension that has $B = 1/2$. More generally we might try

$$B = \frac{1}{2} + n\alpha_2 \tag{57}$$

which leads to the condition

$$n\alpha_1\alpha_2(\rho_1-\rho_2) < (1+B)(\rho_2+\rho_1 B) \tag{58}$$

If $\rho_2 = 0$ (bubbles in a dense liquid), (58) becomes

$$n\alpha_2(1-\alpha_2) < \left(\frac{3}{2}+n\alpha_2\right)\left(\frac{1}{2}+n\alpha_2\right) \tag{59}$$

which can never be violated if n is positive and is only not satisfied over a very narrow range of α_2 for a limited range of negative n.

Steady Incompressible Flow Past Stationary Particles. In this case (7) is valid and we have, from (35)

$$p_1 + \frac{1}{2}\rho_1\alpha_1\beta v_1^2 = \text{constant} \tag{60}$$

as discussed in [4] and [7].

Moreover, from (9) and (10)

$$p_1 - p_2 = -\frac{\alpha_1}{2}\rho_1 v_1^2\frac{d}{dt}(\alpha_1\beta) \tag{61}$$

Combining (60) and (61) we have

$$p_2 - \frac{1}{2}\rho_1 v_1^2 \alpha_1^2 \frac{d\beta}{d\alpha_1} = \text{constant} \tag{62}$$

Using (62) in (7) we have the result derived in [4]

$$\mathbf{f}_2 = \nabla \left(\frac{1}{2}\rho_1 v_1^2 \alpha_1^2 \frac{d\beta}{d\alpha_1} \right) \tag{63}$$

which is compatible with the several examples presented in [4] of forces on particles when various gradients of void fraction or fluid mean velocity are imposed. In particular this should be true in the *dilute limit* where it is customary to assume that c_{vm} in (3) is independent of α_2. From (2) and (3) we have

$$\beta = \frac{(\alpha_2/\alpha_1)c_{vm} + 1}{\alpha_1} \tag{64}$$

Therefore

$$\frac{d\beta}{d\alpha_1} = -\frac{c_{vm} + 1}{\alpha_1^2} - \frac{2\alpha_2 c_{vm}}{\alpha_1^3} \tag{65}$$

The second term in (65) may be neglected in the *dilute limit* and the result substituted in (63) to obtain

$$\mathbf{f}_2 = -(1 + c_{vm})\nabla \left(\frac{1}{2}\rho_1 v_1^2 \right) \tag{66}$$

which is exactly Taylor's result [9] for the force on a particle in a converging flow. We note that conclusions incompatible with (66) would have been obtained if some other definition of $\mathbf{M}_1^d$ had been used that did not vanish when $\mathbf{v}_2 = 0$ (or $\alpha_2 \to 0$).

Multidimensional Steady Flow of Incompressible Phases. In this case the continuity equations reduce to

$$\nabla \cdot (\alpha_1 \mathbf{v}_1) = 0 \tag{67}$$

$$\nabla \cdot (\alpha_2 \mathbf{v}_2) = 0 \tag{68}$$

with the constraint

$$\alpha_1 + \alpha_2 = 1 \tag{69}$$

In addition, if the flow is "potential" and Maxwellian, (37) and (44) become

$$-\nabla\phi = \mathbf{v}_1 - \frac{\alpha_2}{2}(\mathbf{v}_2 - \mathbf{v}_1) \tag{70}$$

$$-\nabla\eta = \mathbf{v}_2 + \frac{\alpha_1}{2}R(\mathbf{v}_2 - \mathbf{v}_1) \tag{71}$$

where

$$R = \rho_1/\rho_2 \tag{72}$$

An additional equation is provided by the condition of compatible phase pressures. For a Maxwellian suspension with a single stagnation pressure, the condition leads to (50).

It would be useful to develop and examine several exact solutions to this set of equations. One very simple case results when the void fraction is uniform throughout the entire flow. Examination of (67) to (71) reveals that $\mathbf{v}_1$ and $\mathbf{v}_2$ satisfy equations identical to those of single phase irrotational incompressible steady flow, i.e.

$$\nabla \cdot \mathbf{v}_i = 0, \ \ \mathbf{v}_i = -\nabla\phi_i \tag{73}$$

If, in addition, the phases have a common stagnation condition, (51) is valid, and (73) may be satisfied if $\mathbf{v}_1$ is the solution to *any* single phase potential flow and

$$\mathbf{v}_2 = \left(\frac{3R}{R+2}\right)^{1/2} \mathbf{v}_1 \tag{74}$$

The phases follow identical streamlines at a constant, uniform slip ratio. This is clearly a possible solution to the "separated flow" regime, with two continuous fluids flowing side by side in alternate streamlines with thicknesses in a constant ratio and $\mathbf{v}_2 = R^{1/2}\mathbf{v}_1$, but does not obviously apply *a priori* to a uniform dispersion.

If the above analysis is valid, there is no "inertial separation" in such a flow. It happens in real fluids because interphase drag reduces the slip ratio so that centrifugal force, for example, has a greater effect on the heavier phase.

One-Dimensional Transients. To develop an equation describing transients we retain $\mathbf{f}_1$ and $\mathbf{f}_2$ in (4) and (5) with the result that these terms add to the pressure gradients in (42) and (47). Eliminating the pressures from (42) and (47) by using (9) and adopting Maxwell's equation (49) we get

$$\rho_2 \frac{\partial}{\partial t}\left(\mathbf{v}_2 + \frac{\alpha_1\rho_1}{2\rho_2}(\mathbf{v}_2 - \mathbf{v}_1)\right) - \rho_1 \frac{\partial}{\partial t}\left(\mathbf{v}_1 - \frac{\alpha_2}{2}(\mathbf{v}_2 - \mathbf{v}_1)\right)$$

$$+\rho_2 \nabla\left(\frac{v_2^2}{2} + \frac{\alpha_1\rho_1}{2\rho_2}(\mathbf{v}_2 - \mathbf{v}_1)\cdot\mathbf{v}_2\right) - \rho_1 \nabla\left(\frac{v_1^2}{2} + \frac{\alpha_2}{4}(v_1^2 - v_2^2)\right)$$

$$-\nabla\left(\frac{\alpha_1}{4}\rho_1(\mathbf{v}_2 - \mathbf{v}_1)\cdot(\mathbf{v}_2 - \mathbf{v}_1)\right) + \mathbf{f}_1 - \mathbf{f}_2 = 0 \qquad (75)$$

If the phases are incompressible, (75) reduces to the surprisingly simple form

$$\left(\rho_2 + \frac{\rho_1}{2}\right)\left(\frac{\partial \mathbf{v}_2}{\partial t} + \frac{\nabla v_2^2}{2}\right) - \frac{3}{2}\rho_1\left(\frac{\partial \mathbf{v}_1}{\partial t} + \frac{\nabla v_1^2}{2}\right) + \mathbf{f}_1 - \mathbf{f}_2 = 0 \qquad (76)$$

In a one-dimensional situation, with motion entirely in the z-direction, (76) becomes

$$\left(\rho_2 + \frac{\rho_1}{2}\right)\left(\frac{\partial v_2}{\partial t} + v_2\frac{\partial v_2}{\partial z}\right) - \frac{3}{2}\rho_1\left(\frac{\partial v_1}{\partial t} + v_1\frac{\partial v_1}{\partial z}\right) + f_1 - f_2 = 0 \qquad (77)$$

(77) has exactly the same form as would be obtained from a one-dimensional analysis that ignored the effects of added mass (e.g. (6.137) of [10]) except that the effective densities are now different. The continuity equations (26) and (27) reduce, for incompressible phases, to

$$\frac{\partial \alpha_1}{\partial t} + \alpha_1\frac{\partial v_1}{\partial z} + v_1\frac{\partial \alpha_1}{\partial z} = 0 \qquad (78)$$

$$\frac{\partial \alpha_2}{\partial t} + \alpha_2\frac{\partial v_2}{\partial z} + v_2\frac{\partial \alpha_2}{\partial z} = 0 \qquad (79)$$

and the compatibility condition (15) provides the fourth equation for solution in terms of the four variables α_1, α_2, v_1, v_2. Since these equations have exactly the same form as in the case where added mass is ignored, there is no change in the form of the solutions, such as those derived in [10], nor is there any change in possible problems with unreal characteristics when the relative velocity is large or when $f_1 - f_2$ contains no influence of gradients or higher order differentials.

Summary. Geurst's equations are consistent with several specific results derived in [4]. A new conclusion is that, when Maxwell's expression for the *exertia* (representing inertial coupling) is used, the one-dimensional transient equations for a pair of incompressible phases do not change in character when inertial coupling effects are included. The only effect is to change the effective densities of the phases. Thus, one-dimensional characteristics of unsteady flow in fluidized systems are not altered in any fundamental way when inertial coupling (added mass) effects are included.

REFERENCES

1. Geurst, J.A., Virtual Mass in Two-Phase Bubbly Flow, *Physica, 129A*, 233, 1985.
2. Geurst, J.A., Two-Fluid Hydrodynamics of Bubbly Liquid/Vapour Mixture Including Phase Change, *Philips J. Res., 40*, 352, 1985.
3. Geurst, J.A., Variational Principles and Two-Fluid Hydrodynamics of Bubbly Liquid/Gas Mixtures, *Physica, 135A*, 455, 1986.
4. Wallis, G.B., Inertial Coupling in Two-Phase Flow: Macroscopic Properties of Suspensions in an Inviscid Fluid, accepted for publication by *Multiphase Science & Technology*, 1989.
5. Drew, D.A. and Wood, R.T., Overview and Taxonomy of Models and Methods, *International Workshop on Two-Phase Flow Fundamentals*, National Bureau of Standards, Maryland, September, 1985.
6. Drew, D.A. and Lahey, R.T., Jr., The Virtual Mass and Lift Force on a Sphere in Rotating and Straining Inviscid Flow, *Int. J. Multiphase Flow, 13*, 113, 1987.
7. Wallis, G.B. Some Tests of Two-Fluid Models for Two-Phase Flow, presented at the U.S.-Japan Seminar on Two-Phase Flow Dynamics, Ohtsu, Japan, 1988.
8. Pauchon, C. and Banerjee, S., Interphase Momentum Interaction Effects in the Averaged Multifluid Model, Part 1, Void Propagation in Bubbly Flow, *Int. J. Multiphase Flow, 12*, 559, 1986.
9. Taylor, G.I., The Forces on a Body Placed in a Curved or Converging Stream of Fluid, *Proc. Roy. Soc., A120*, 260-283, 1928.
10. Wallis, G.B., *One-Dimensional Two-Phase Flow*, McGraw-Hill, New York, 1969.

MIX
Papier aus verantwortungsvollen Quellen
Paper from responsible sources
FSC® C105338

If you have any concerns about our products,
you can contact us on
ProductSafety@springernature.com

In case Publisher is established outside the EU,
the EU authorized representative is:
Springer Nature Customer Service Center GmbH
Europaplatz 3, 69115 Heidelberg, Germany

Printed by Libri Plureos GmbH
in Hamburg, Germany